Edexcel
GCSE Mathematics

higher

practice book

Keith Pledger

Gareth Cole

Peter Jolly

Graham Newman

Joe Petran

www.heinemann.co.uk
✓ Free online support
✓ Useful weblinks
✓ 24 hour online ordering

01865 888058

Heinemann

Inspiring generations

Heinemann Educational Publishers
Halley Court, Jordan Hill, Oxford OX2 8EJ
Part of Harcourt Education Limited

Heinemann is the registered trademark of
Harcourt Education Limited

© Harcourt Education Ltd, 2006

First published 2006

10 09 08 07 06
10 9 8 7 6 5 4 3 2

British Library Cataloguing in Publication Data is available from the British Library on request.

10-digit ISBN: 0 435 53363 0
13-digit ISBN: 978 0 435533 63 2

10-digit ISBN (10-pack): 0 435 53378 9
13-digit ISBN (10-pack): 978 0 435533 78 6

Typeset by Tech-Set Ltd, Gateshead, Tyne and Wear
Original illustrations © Harcourt Education Limited, 2006
Illustrated by Beehive
Cover design by mccdesign
Printed and bound at Scotprint, Haddington, Scotland
Cover photo: PhotoLibrary.com

Acknowledgements
Harcourt Education Ltd would like to thank those schools who helped in the development and trialling of this course.

This high quality material is endorsed by Edexcel and has been through a rigorous quality assurance programme to ensure that it is a suitable companion to the specification for both learners and teachers. This does not mean that its contents will be used verbatim when setting examinations nor is it to be read as being the official specification – a copy of which is available at www.edexcel.org.uk

The publisher's and authors' thanks are due to Edexcel Limited for permission to reproduce questions from past examination papers. These are marked with an [E]. The answers have been provided by the authors and are not the responsibility of Edexcel Limited.

The authors and publisher would like to thank the following individuals and organisations for permission to reproduce photographs:

Tudor Photography p19; iStockPhoto p22; Alamy Images pp36, 38, 101, 107; Photos.com p54; Science Photo Library p56; Getty Images p105

Every effort has been made to contact copyright holders of material reproduced in this book. Any omissions will be rectified in subsequent printings if notice is given to the publishers.

Tel: 01865 888058   www.heinemann.co.uk

Contents

1	Exploring numbers 1	1
2	Essential algebra	6
3	Shapes	10
4	Fractions and decimals	16
5	Collecting and recording data	19
6	Solving equations and inequalities	25
7	Transformations and loci	29
8	Using basic number skills	35
9	Functions, lines, simultaneous equations and regions	42
10	Presenting and analysing data 1	48
11	Estimation and approximation	54
12	Sequences and formulae	57
13	Measure and mensuration	60
14	Simplifying algebraic expressions	64
15	Pythagoras' theorem	68
16	Basic trigonometry	73
17	Graphs and equations	77
18	Proportion	82
19	Quadratic equations	86
20	Presenting and analysing data 2	90
21	Advanced trigonometry	95
22	Advanced mensuration	100
23	Exploring numbers 2	106
24	Probability	110
25	Transformations of graphs	117
26	Circle theorems	121
27	Vectors	128
28	Introducing modelling	132
29	Conditional probability	136

About this book

This revised and updated edition provides a substantial number of additional exercises to complement those in the Edexcel GCSE Mathematics higher textbook. Extra exercises are included for almost every topic in the course textbook.

The author team is made up of Senior Examiners, a Chair of Examiners and Senior Moderators, all experienced teachers with an excellent understanding of the Edexcel specification.

Clear links to the course textbook help you plan your use of the book.

Please note that the answers to the questions are provided in a separate booklet. This will be supplied with each order for 10-packs, or with your first order for single books.

1 Exploring numbers 1

Exercise 1.1   Link 1A

1 Find all the factors of
   (a) 12   **(b)** 37   **(c)** 72
   (d) 616   **(e)** 49   **(f)** 2002

> **Remember:** A factor of a number is a whole number that divides exactly into the number.

2 Work out the sum of the prime numbers between 20 and 30.

3 Which of these numbers are prime?
   (a) 3   **(b)** 23   **(c)** 27
   (d) 1   **(e)** 91   **(f)** 97

> **Remember:** A prime number is a number greater than 1 which has only two factors – itself and 1.

4 Work out the value of
   (a) $2^3 \times 3^2 \times 5$ **(b)** $3^3 \times 5^2$
   (c) $2^5 \times 3^2$ **(d)** $2^2 \times 5^2 \times 11^3$

5 Given that $2^n \times 3 = 24$, work out the value of n.

6 The number 1998 can be written as $2 \times 3^n \times p$, where n is a whole number and p is a prime number.
   (a) Work out the values of n and p.
   (b) Using your answers to part **(a)**, or otherwise, work out the factor of 1998 which is between 100 and 200.   [E]

7 Write each of these numbers in prime factor form.
   (a) 36   **(b)** 72   **(c)** 39
   (d) 1440   **(e)** 200   **(f)** 2002

> You may find using a factor tree is a good method to work out this.

8 Find the highest common factor (HCF) of
   (a) 6 and 15   **(b)** 48 and 72
   (c) 1500 and 504   **(d)** 99 and 3003

> **Remember:** The HCF of two whole numbers is the highest factor that is common to both of them.

9 Find the HCF of
   (a) 32 and 80
   (b) 24 and 117
   (c) 72 and 1960

10 Find the lowest common multiple (LCM) of
   (a) 24 and 30   **(b)** 48 and 108
   (c) 60 and 45   **(d)** 36 and 49

> **Remember:** The LCM of two numbers is the lowest number that is a multiple of them both.

11 Two numbers, n and m, have a LCM of $n \times m$. What can you say about n and m?

12 Mr Khan has two flashing lamps.
He switches both of them on at exactly the same time.
The first lamp flashes every 25 seconds.
The second lamp flashes every minute.
Work out the first three times after the
lamps are switched on when they will flash together.

Exercise 1.2   Link 1B

1 Copy and complete this table of values for triangular numbers.

1st	2nd	3rd	4th	5th	6th	7th	8th	9th	10th
1	3	6	10	15	21	28	36	45	55

2 Copy and complete this table of values for square numbers.

1st	2nd	3rd	4th	5th	6th	7th	8th	9th	10th
1	4	9	16	25	36	49	64	81	100

3 Copy and complete this table of values for cube numbers.

1st	2nd	3rd	4th	5th	6th	7th	8th	9th	10th
1	8	27	64	125	216	343	512	729	1000

4 By considering the dot patterns, show that the sum of two consecutive triangular numbers is a square number.

5 (a) Show that 25 and 144 are square numbers.

(b) Show that $25 + 144$ is also a square number.

(c) Find two other square numbers, both of which must be greater than 25, which add together to give another square number.

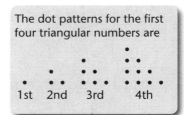

The dot patterns for the first four triangular numbers are

1st   2nd   3rd   4th

6 The 3rd square number is 9.
The 4th square number is 16.
The difference between these two consecutive square numbers is
$16 - 9 = 7$.
This result, 7, is an odd number.
By considering dot patterns for square numbers, show that the difference between any two consecutive square numbers is an odd number.

The dot patterns for the first three square numbers are

1st   2nd   3rd

7 The notation C_4 stands for the 4th cube number; so
$C_4 = 4^3 = 4 \times 4 \times 4 = 64$. Show that $C_3 + C_4 + C_5 = C_6$.

8 $1 = 1$
    $1 + 3 = 4$
    $1 + 3 + 5 = 9$
    $1 + 3 + 5 + 7 = 16$

    (a)   What do you notice is happening with this pattern of results?

    (b)   Use this to find the sum of the first 20 odd numbers.

9 $1 = 1$
    $3 + 5 = 8$
    $7 + 9 + 11 = 27$
    $13 + 15 + 17 + 19 = 64$

    (a)   What do you notice is happening with this pattern of results?

    (b)   Write 1000 as the sum of consecutive odd numbers.

10   Find the next two numbers in the sequence

       1, 9, 36, 100, 225, ...

> Use cube numbers.

11   Prove that the difference between the squares of any two consecutive odd numbers is a multiple of 8.

12   The odd numbers, starting at 1, are set out in triangles.
The 4th and 5th triangles of odd numbers are shown on the right.
The sum of the numbers in the bottom row of the 4th triangle of odd numbers is

       $13 + 15 + 17 + 19 = 64$

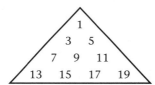

4th triangle of odd numbers

    (a)   Write down, in terms of n, an expression for the sum of the numbers in the bottom row of the nth triangle of odd numbers.

    (b)   Use your expression to find the sum of the numbers in the bottom row of the 50th triangle of odd numbers.

    (c)   The sum of the numbers in the bottom row of the nth triangle of odd numbers is 3375.
Find the value of n.

    (d)   Investigate triangles of odd numbers to find an expression for the sum of all the numbers in the nth triangle of odd numbers.

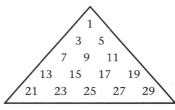

5th triangle of odd numbers

    (e)   Prove that
       $(1 + 2 + 3 + 4 + \cdots + n)^2 = 1^3 + 2^3 + 3^3 + 4^3 + \cdots + n^3$ [E]

Exercise 1.3   **Links 1C–F**

1   Calculate

    (a) 3^3
       (b) 5^2
       (c) 10^3

    (d) $7^2 \times 2^3$
       (e) 4^2
       (f) 3^2

    (g) 7^5
       (h) 12^3
       (i) $3^2 \times 2^3$

    (j) $\sqrt{169}$
       (k) $(8.1)^2$
       (l) 53.2^2

    (m) $\sqrt[3]{72}$
      (n) $\sqrt[3]{-15.2}$
     (o) $\sqrt[3]{7457.14}$

> **Remember:**
> $3^3 = 3 \times 3 \times 3$

> Use your calculator for parts **(k)** to **(o)**.

2 Write these numbers in index form.

(a) 10 000   (b) 1 000 000   (c) 225

(d) 343   (e) 289   (f) 1024

(g) 2197   (h) 625

3 Which is the greater, and by how much:

(a) 2^2 or 3^2 (b) 3^4 or 4^3?

4 Work out each of the following.
In each case leave your answers both in index form and, where possible, without using indices.

(a) $7^2 \times 7^3$ (b) $10^3 \times 10^2$ (c) $2^5 \times 2^3$

(d) $3^3 \times 3^2 \times 3$ (e) 12^0 (f) $5^3 \div 5^2$

(g) $7^5 \div 7^3$ (h) $17^3 \times 17^0$ (i) $8^3 \times 4^2$

(j) $7^2 \div 7^3$ (k) $15^3 \div 15$ (l) $15 \div 15^3$

(m) $3 + 3^2 + 3^2$ (n) 2^{-1} (o) $(3^2)^3$

(p) $(3^3)^2$ (q) $(7^2)^2$ (r) 10^{-2}

(s) $(5^{-1})^2$ (t) $5^3 \div 5^{-2}$ (u) $(2^2)^{-3}$

(v) $10^{-2} \div 10^{-3}$ (w) $(2^{-2})^{-3}$ (x) $5^3 \times 5^2 \div 5^{-2}$

> **Remember:** The rules of indices are
> $a^n \times a^m = a^{n+m}$
> $a^n \div a^m = a^{n-m}$
> $a^0 = 1$
> $a^{-n} = \dfrac{1}{a^n}$
> $(a^n)^m = a^{n \times m}$

5 Work out each of these, leaving your answer as an ordinary number.

(a) $\dfrac{2^2 \times 2^3}{2}$ (b) $\dfrac{3^3 \times 3^2}{3^{-1}}$ (c) $\dfrac{(5^2)^3}{5^{-2}}$ (d) $\dfrac{(4^2)^3}{4^7}$

6 Without using a calculator, work out the exact value of

(a) $16^{\frac{1}{2}}$ (b) $10\,000^{\frac{1}{4}}$ (c) $32^{\frac{1}{5}}$

(d) $64^{\frac{1}{3}}$ (e) $64^{-\frac{1}{3}}$ (f) $1000^{-\frac{1}{3}}$

(g) $25^{\frac{3}{2}}$ (h) $49^{-\frac{3}{2}}$ (i) $169^{\frac{1}{2}}$

(j) $169^{-\frac{1}{2}}$ (k) $169^{\frac{3}{2}}$ (l) $169^{-\frac{3}{2}}$

(m) $(4^3)^{-\frac{2}{3}}$ (n) $1024^{-\frac{1}{10}}$ (o) $225^{-\frac{1}{2}}$

(p) $8^{\frac{4}{3}}$ (q) $8^{-\frac{4}{3}}$ (r) $8^{\frac{1}{3}} \times 2^{-1}$

(s) $27^{\frac{5}{3}}$ (t) $2^{\frac{1}{3}} \times 4^{\frac{1}{3}}$ (u) $25^{-\frac{3}{2}}$

(v) $32^{-\frac{3}{5}}$ (w) $(4^2)^{-\frac{1}{4}}$ (x) $125^{\frac{4}{3}}$

(y) $1\,000\,000^{-\frac{2}{3}}$

> $a^{\frac{1}{n}} = \sqrt[n]{a}$

7 Given that $8^{\frac{1}{3}} \times p^n \times 21 = 1050$, where p is a prime number and n is a positive integer, find

(a) the value of p (b) the value of n.

8 Express as a single fraction:

(a) $(2^0 + 2^{-1} + 2^{-2} + 2^{-3})^2$ (b) $(2^0 + 2^{-1} + 2^{-2} + 2^{-3})^{-1}$

9 **(a)** Show that
   (i) $1^2 = 1$ **(ii)** $11^2 = 121$ **(iii)** $111^2 = 12\,321$

   (b) Work out the value of
   (i) 1111^2 **(ii)** $11\,111^2$

   (c) By looking at the pattern of the results in parts **(a)** and **(b)**, write down the value of
   (i) $111\,111^2$ **(ii)** $111\,111\,111^2$

Exercise 1.4   **Link 1G**

1 Work out
   (a) 2^6 **(b)** 2^{-6} **(c)** 2^{-3}
   (d) 10^3 **(e)** 10^{-3} **(f)** $2^3 \times 10^4$

2 Round 2^9 to the nearest 10.

3 Solve
   (a) $2^{3n+1} = 32$ **(b)** $10^{5-2n} = 10\,000\,000$

4 Find an expression for the sum of the series
$$10 + 10^2 + 10^3 + 10^4 + \cdots + 10^n$$

② Essential algebra

1 Work out the value of these algebraic expressions using the values given.

(a) $5a + 3$ if $a = 4$

(b) $4b - c$ if $b = 2$, $c = 5$

(c) $3p - 2q$ if $p = 5$, $q = 2$

(d) $xy - z$ if $x = 2$, $y = 4$, $z = 3$

(e) $12 + 5t$ if $t = -2$

(f) $p - 3t$ if $p = 4$, $t = -2$

(g) $4y + 7$ if $y = 4\frac{1}{2}$

(h) $6st$ if $s = \frac{1}{2}$, $t = \frac{3}{4}$

(i) $4(a + b)$ if $a = 2$, $b = 3$

(j) $5(x - y)$ if $x = 7$, $y = 4$

(k) $x(6 - y)$ if $x = 3$, $y = 2$

(l) $3(8 - t)$ if $t = -2$

(m) $\frac{1}{2}(a + b)$ if $a = 3$, $b = 5$

(n) $-2(3t + 1)$ if $t = -2$

(o) $3(2x - y)$ if $x = -1$, $y = -3$

(p) $4(p + 2q)$ if $p = 1$, $q = -2\frac{1}{2}$

(q) $6(a + b)$ if $a = 2$, $b = -2$

(r) $\frac{3}{4}(f + g)$ if $f = 2$, $g = -10$

> Use BIDMAS to help you remember the order of operations.
> Follow this order:
>   Brackets
>   Indices
>   Division
>   Multiply
>   Add
>   Subtract

2 Work out the value of each of these expressions for the given values of the letters.

(a) $\frac{x}{5} + 2$ if $x = 15$

(b) $\frac{p}{q} - 2$ if $p = 36$, $q = 3$

(c) $\frac{n - 3}{5}$ if $n = 13$

(d) $\frac{a + 2b}{3}$ if $a = 1$, $b = 4$

(e) $\frac{ab + c}{2}$ if $a = 2$, $b = 3$, $c = 4$

(f) $\frac{2pq}{12} - 3$ if $p = 4$, $q = 6$

(g) $\frac{4xy - z}{7}$ if $x = 2$, $y = 3$, $z = -4$

(h) $\frac{3t}{15} + s$ if $t = 5$, $s = -1$

(i) $2t - \frac{3r}{4}$ if $t = 5$, $r = 8$

(j) $a - \frac{4b}{6}$ if $a = -2$, $b = -12$

> **Remember:** In the expression $\frac{p}{q} - 2$, the line acts as a bracket. Work out this first.

3 Work out the value of each of these algebraic expressions using the values given.

(a) $3t^2 + 2$ if $t = 4$

(b) $2n^2 - 3$ if $n = 5$

(c) $4x^2$ if $x = \frac{1}{2}$

(d) $8y^2 + 1$ if $y = -1$

(e) $5 - p^2$ if $p = -3$

(f) $3a^2 + 2b^2$ if $a = -1$, $b = 3$

(g) $x^2 + 2x$ if $x = 4$

(h) $3t^2 - 2t$ if $t = 5$

(i) $x^2 - y^2$ if $x = 10$, $y = 3$

(j) $s^2 - t^2$ if $s = -8$, $t = 6$

(k) $\frac{1}{2}(x + x^2)$ if $x = 5$

(l) $mn + m + n$ if $m = n = 4$

(m) $\frac{x^2}{2} - 3$ if $x = 5$

(n) πr^2 if $\pi = \frac{22}{7}$, $r = 7$

(o) $\frac{a^2}{3} - \frac{b^2}{8}$ if $a = 6$, $b = 4$

(p) $3a(x^2 + y^2)$ if $a = 2$, $x = 3$, $y = 4$

> **Remember:** In the expression $3t^2 + 2$, work out t^2 first, as indices come first in the order of operations.

(q) $\dfrac{2v^2 + u^2}{10}$ if $v = 5$, $u = 3$

(r) $\frac{1}{2}(a^2 - b)$ if $a = -4$, $b = -6$

(s) $\dfrac{a - 3b^2}{4}$ if $a = -1$, $b = 1$

(t) $\dfrac{x^2 - 3y^2}{4z^2}$ if $x = -5$, $y = -3$, $z = -\frac{1}{2}$

Exercise 2.2   Links 2C–F

1 Simplify

(a) $x^3 \times x^2$	**(b)** $y^4 \times y^5$	**(c)** $a^3 \times a^5$
(d) $a \times a^2$	**(e)** $(b^2)^3$	**(f)** $x^6 \div x^2$
(g) $b^8 \div b^3$	**(h)** $c^{20} \div c^{14}$	**(i)** $n^2 \div n$
(j) $(b^4)^5$	**(k)** $(a^3)^0$	**(l)** $x^3 \times x^2 \times x$
(m) $4x^3 \times 3x^2$	**(n)** $5p \times 2p^3$	**(o)** $2y^{10} \times y^3$
(p) $6x^3 \div 2x$	**(q)** $24p^7 \div 8p^3$	**(r)** $12a^6 \div 3a^2$
(s) $(3a^2)^3$	**(t)** $(5b^7)^2$	**(u)** $\dfrac{4x^3 \times 6x^2}{3x^4}$
(v) $\dfrac{a^4 \times a^3}{a^2}$	**(w)** $\dfrac{3b^3 \times 4b}{2b^4}$	**(x)** $\dfrac{(4a^5)^2}{8a^3}$

$x^3 \times x^2 = x \times x \times x \times x \times x$

$\dfrac{4x^3 \times 6x^2}{3x^4}$

$= \dfrac{4 \times 6 \times x \times x \times x \times x \times x}{3 \times x \times x \times x \times x}$

2 Expand the brackets.

(a) $3(x + 2)$	**(b)** $5(2p + 1)$	**(c)** $4(x - 3)$
(d) $2(3p - 2)$	**(e)** $3(a + 2b)$	**(f)** $9(2n - 5)$
(g) $x(2x + 5)$	**(h)** $a(x + a)$	**(i)** $p(3p + 1)$
(j) $3x(2x - 1)$	**(k)** $n(n + 2)$	**(l)** $4b(b + 2)$
(m) $2t^2(t + 3)$	**(n)** $5x(2 - x)$	**(o)** $4y^3(2y - 3y^2)$

Remember: Multiply each term inside the bracket by the term outside.

3 Simplify each of these expressions.

(a) $3(2x + 1) + x$	**(b)** $5(2a + 1) - 3a$
(c) $2(x + 3y) + x + y$	**(d)** $3(3p - q) + 4(p + 2q)$
(e) $4(3a - 2b) + 2(a + b)$	**(f)** $x(x - 2) + x$
(g) $y(2y - 3) - 5y$	**(h)** $5n(2 + 3n) - 6n$
(i) $m(m^2 + m) - 2m^2$	**(j)** $4t^2(t - 1) - t^3$
(k) $\frac{1}{2}(2x^2 + 4x) + x^2 - x$	**(l)** $3n^2(n^3 + 2n) + n(n^2 - 3n)$
(m) $5x(2x + x^3) - x^4$	**(n)** $y^2(3y^2 + 5y) + 2y^3(1 - 3y)$

Expand the brackets first, then collect like terms.

4 Simplify each of these expressions.

(a) $\dfrac{3n + 12}{3}$ (b) $\dfrac{15x - 10}{5}$ (c) $\dfrac{2a(3a + 6)}{2}$

(d) $\dfrac{6x^2 - 12x}{3}$ (e) $\dfrac{8r + 20s}{4}$ (f) $\dfrac{12a + 6b + 9c}{3}$

(g) $\dfrac{49x - 42y}{7}$ (h) $\dfrac{6a^2 + 4a^2}{5a}$ (i) $\dfrac{5(6t + 8s)}{10}$

> The line acts as a bracket, so work out the 'top' first.

5 Simplify each of these expressions.

(a) $5x - 2(x - y)$ (b) $8t - 3(2t - 1)$

(c) $3(2n + m) - 5(n - m)$ (d) $2(3x - 2) - 4(x - 1)$

(e) $4(3 - x) - (3 + x)$ (f) $a(a - b) - a(a + b)$

(g) $6x(x + 3) - x(6 + x)$ (h) $\frac{1}{2}(4x^2 + 6) - (1 - x^2)$

(i) $\dfrac{4x + 6}{2} - (x - 2)$ (j) $5y - 1 - \dfrac{6y - 2}{2}$

(k) $3(2 - a) - \dfrac{4a - 10}{2}$ (l) $p(p - 1) - p(1 - p)$ (m) $x(3x - 4) - 2(2x^2 - 5)$

(n) $-3(b - a) - (a + b)$ (o) $p^3\left(3 - \dfrac{1}{p}\right) - \frac{1}{2}(p^2 - 2)$ (p) $x(x - y) - y(x + y)$

(q) $x(5x + 4) - 2x(x - 3)$ (r) $n^2(1 - n) - n(1 + n)$ (s) $y^3(1 + y) - y(3 + y^3)$

(t) $8t(2 - \frac{1}{2}t) + 3(1 + t^2)$ (u) $3x - 3(x - y)$ (v) $4a(1 - a^2) - 3(a + a^2)$

> **Remember:** Use BIDMAS to help you with the order of operations.

Exercise 2.3   Links 2G–I

1 Factorise completely.

(a) $4x + 20$ (b) $2y + 16$ (c) $3a + ab$

(d) $xy - 2y$ (e) $3z^2 + 9$ (f) $4x + 8x^2$

(g) $2ax - axy$ (h) $7z^2 + 21z$ (i) $ay + ax$

(j) $15ab - 5ac$ (k) $12ab^2 - 3ab$ (l) $8x^2y^2 - 2xy$

> **Remember:** The highest common factor must appear outside the brackets.

2 Expand and simplify.

(a) $(x + 3)(x + 2)$ (b) $(x + 7)(x - 3)$

(c) $(x - 4)(x - 5)$ (d) $(x + 1)(x - 5)$

(e) $(x + 7)^2$ (f) $(2 - x)(x + 4)$

(g) $(3 - 2x)(4 - 3x)$ (h) $(5x + 2)(2 - 5x)$

(i) $(x + 3)(x - 3)$ (j) $(10 - 2x)(10 + 2x)$

(k) $(3x + 1)(2x - 2)$ (l) $(4x - 3)(2x - 7)$

(m) $(3 + 5x)(2x - 1)$ (n) $(7x - 3)(2 + 3x)$

(o) $(2x - 3)(2x + 3)$ (p) $(7 - 3x)(7 + 3x)$

(q) $(y + b)^2$ (r) $(5 - x)^2$

(s) $(x - y)^2$ (t) $(8 + x)^2$

> You may wish to draw a rectangle to help you.

3   Multiply out the brackets and simplify.

(a) $(p + 2q)(q + 2p)$ (b) $(x + y)(x - y)$
(c) $(x - y)(x + 2y)$ (d) $(n - 2m)(n + 3m)$
(e) $(a + 3b)(3a - b)$ (f) $(2x + 3y)(3x + 4y)$
(g) $(4a - 1)(a + 2)$ (h) $(3x - 2)(4x + 3)$
(i) $(x + 5)^2$ (j) $(y + 1)^2$
(k) $(2a + 3)^2$ (l) $(4x - 3)^2$
(m) $(2 - x)^2$ (n) $(3 - 5y)^2$
(o) $(a - 3)^2$ (p) $(x + y)^2$
(q) $(3x + 2y)^2$ (r) $(4p - 5y)^2$
(s) $(2p + 3)(p + 2) - 2p^2$ (t) $(n - 3)(n + 1) + 2n$

Exercise 2.4   Link 2J

Solve the equations.

1 $2x + 1 = 9$	**2** $3x + 2 = 20$	**3** $4x = 24$
4 $8 + 5y = 18$	**5** $3b - 6 = 21$	**6** $7c - 8 = 20$
7 $9k + 1 = 26$	**8** $3k - 17 = -4$	**9** $5k + 11 = -20$
10 $2 - 3k = 17$	**11** $4 - 5m = 22$	**12** $3 = 7 - m$
13 $12 = 2p + 7$	**14** $17 = 18 - 5p$	**15** $2 - q = 19$

3 Shapes

1 Calculate the size of the lettered angles.

> **Remember:**
> Angles on a straight line
> Angles at a point
> Alternate angles
> Corresponding angles
> Co-interior angles

(a)

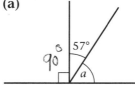

(b)

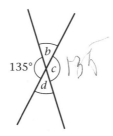

(c)

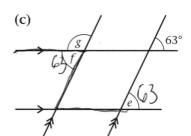

(d)

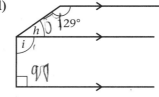

2 Calculate the size of the angles marked with letters.
Write down what type each triangle is.

(a)

(b)

(c)

(d)

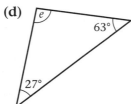

(e)

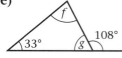

(f)

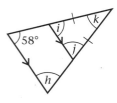

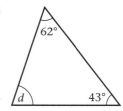

Exercise 3.2

1 Work out the lettered angles in these shapes.
Write down your reasons.

(a)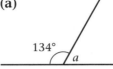

134°
a

(b)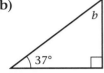

b
37°

(c)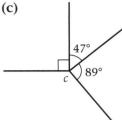

47°
c 89°

(d)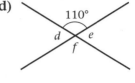

110°
d e
f

(e)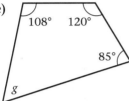

108° 120°
85°
g

(f)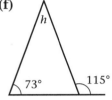

h
73° 115°

2 Draw a copy of each diagram and find all the unmarked angles.
Write down the reasons.

(a)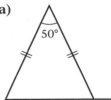

50°

(b)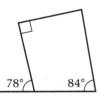

75°

(c)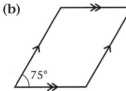

78° 84°

(d)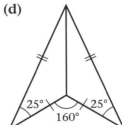

25° 25°
160°

(e)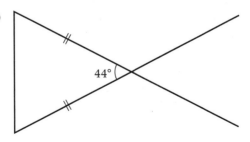

44°

3 Calculate the size of the interior angles of a regular hexagon.

4 Calculate the unknown angles in these polygons.

Exterior angles are useful.

(a)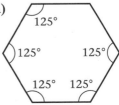

125°
125° 125°
125° 125°

(b)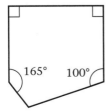

165° 100°

(c)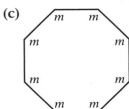

m m
m m
m m
m m

5 This diagram shows a regular octagon.
Calculate the value of *x* and *y*.

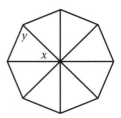

Exercise 3.3   **Link 3E**

1 Sketch a plan, front elevation and side elevation for each of these
solids.

(a) 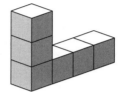   **(b)**

2 Sketch a net to make an octagonal pyramid.

3 Sketch a net to make this solid.

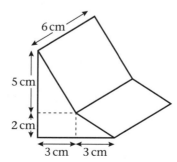

> Imagine the shape
> unfolded.

4 The diagram shows a scale drawing. Draw the plan and the front
and side elevations to scale on squared paper.

Exercise 3.4

1 Which of these pairs of triangles are congruent? List the vertices in corresponding order and give reasons for congruency.

(a) (i)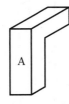

(ii)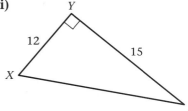

> **Remember:**
> S S S
> S A S
> R H S
> A S A

(b) (i)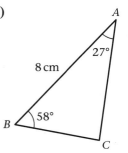

(ii)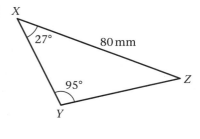

(c) (i)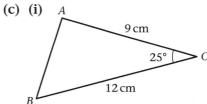

(ii)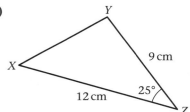

2 State which pairs of shapes are congruent.

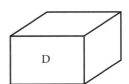
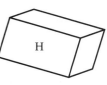

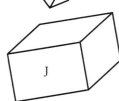

3 **(a)** *ABCD* is a rhombus.
Prove that *ABC* is congruent
to *ADC*.

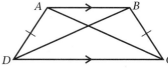

(b)

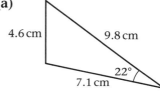

In the diagram
AD = *BC*
∠*ADC* = ∠*BCD*

Prove that triangles *CAD* and *DBC* are congruent.

Exercise 3.5   Link 3G

1 Write down why these pairs of triangles are similar.
Calculate the length of each side marked by a letter.

(a)

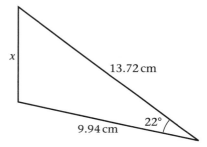

(b)

Draw the triangles so they both face the same way.

(c)

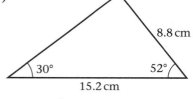

2 **(a)** Explain why these two triangles are similar.

(b) Calculate the lengths p and q.

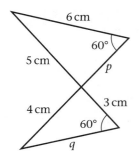

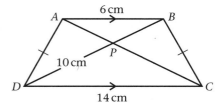

3 *ABCD* is an isosceles trapezium.

(a) Name the similar triangles.

(b) Explain why they are similar.

(c) Calculate the lengths
AP and *PC*.

Exercise 3.6   **Link 3H**

1 Copy each shape and show *all* planes of symmetry.
Write down how many planes of symmetry each shape has.

2 Write down the number of planes of symmetry for a cube.

4 Fractions and decimals

Exercise 4.1

Links 4A, B

1 Copy and complete the equivalent fractions.

(a) $\dfrac{3}{9} = \dfrac{}{63} = \dfrac{6}{} = \dfrac{18}{} = \dfrac{}{27}$

(b) $\dfrac{2}{5} = \dfrac{}{15} = \dfrac{}{30} = \dfrac{14}{} = \dfrac{18}{}$

(c) $\dfrac{7}{9} = \dfrac{}{27} = \dfrac{28}{} = \dfrac{}{81} = \dfrac{49}{}$

(d) $\dfrac{3}{} = \dfrac{15}{20} = \dfrac{}{40} = \dfrac{45}{} = \dfrac{}{100}$

2 Cancel the following fractions to their simplest form.

(a) $\dfrac{44}{66}$ (b) $\dfrac{24}{80}$ (c) $\dfrac{16}{96}$ (d) $\dfrac{25}{85}$ (e) $\dfrac{42}{105}$

3 Arrange in order of size. Start with the smallest.

> Write the fractions with the same denominator.

(a) $\dfrac{5}{8}, \dfrac{1}{2}, \dfrac{11}{16}, \dfrac{3}{4}$ (b) $\dfrac{3}{10}, \dfrac{29}{100}, \dfrac{2}{5}$ (c) $\dfrac{2}{6}, \dfrac{7}{18}, \dfrac{4}{9}$

(d) $\dfrac{2}{7}, \dfrac{13}{42}, \dfrac{7}{21}$ (e) $\dfrac{3}{4}, \dfrac{8}{12}, \dfrac{11}{16}$ (f) $\dfrac{3}{8}, \dfrac{5}{12}, \dfrac{11}{30}$

(g) $\dfrac{9}{14}, \dfrac{13}{21}, \dfrac{23}{35}$

Exercise 4.2

Links 4C–E

1 Rearrange the measurements in each part in order of size. Put the smallest first.

(a) 3.1 cm, 3.12 cm, 3.102 cm, 3.012 cm

(b) 18.32 kg, 18.326 kg, 18.302 kg, 18.236 kg

(c) 9.09 s, 9.089 s, 9.0901 s, 9.101 s

(d) 1.03 ml, 1.1 ml, 1.009 ml, 1.101 ml

(e) 15.3°, 15.27°, 15.208°, 15.268°

2 Work out

(a) $16.32 - 5.11$ (b) $108.5 - 79.6$ (c) $10.8 + 9.3$

(d) $26.5 \div 5$ (e) $18.4 \div 4$ (f) $111.9 \div 3$

(g) 2.05×2 (h) 4.17×3 (i) 2.89×4

(j) 10.109×5

3 Work out

  (a) $2.74 \div 0.2$ **(b)** $0.28 \div 0.4$ **(c)** 1.4×0.3
  (d) 23.4×0.5 **(e)** 0.8×0.3 **(f)** $0.064 \div 0.8$
  (g) 0.07×0.3 **(h)** $300 \div 1.5$ **(i)** $30 \div 0.15$
  (j) $3000 \div 0.015$

 4 Simplify

  (a) $\dfrac{15.5}{0.5}$ **(b)** $\dfrac{12.3}{3}$ **(c)** $\dfrac{4.076}{0.4}$ **(d)** $\dfrac{0.027}{0.003}$

5 Work out

  (a) 23.4×0.35 **(b)** 0.026×0.825 **(c)** $(0.3)^3$

6 Work out to three significant figures

  (a) $13.6 \div 0.023$ **(b)** $0.728 \div 0.0017$ **(c)** $1 \div (0.2)^3$

> You are expected to deal with
> 3 digits $\times$ or $\div$
> 2 digits without a calculator.

Exercise 4.3   Links 4F–H

1 Change these fractions into decimals.

  (a) $\frac{2}{5}$ **(b)** $1\frac{3}{4}$ **(c)** $\frac{7}{8}$ **(d)** $\frac{7}{20}$ **(e)** $\frac{3}{50}$
  (f) $1\frac{1}{2}$ **(g)** $3\frac{9}{20}$ **(h)** $2\frac{2}{25}$ **(i)** $15\frac{3}{50}$ **(j)** $7\frac{7}{200}$

2 Change these decimals into fractions.

  (a) 0.35 **(b)** 0.42 **(c)** 0.6 **(d)** 0.16 **(e)** 0.64
  (f) 0.625 **(g)** 0.85 **(h)** 0.86 **(i)** 1.05 **(j)** 3.84

3 Convert these fractions to recurring decimals.

  (a) $\frac{1}{6}$ **(b)** $\frac{2}{9}$ **(c)** $\frac{3}{11}$ **(d)** $\frac{5}{7}$ **(e)** $\frac{7}{99}$
  (f) $\frac{3}{14}$ **(g)** $\frac{2}{21}$ **(h)** $\frac{7}{48}$ **(i)** $\frac{19}{22}$ **(j)** $\frac{8}{13}$

Exercise 4.4   Links 4I, J

1 Work out

  (a) $\frac{3}{8} + \frac{3}{4}$ **(b)** $\frac{4}{7} + \frac{2}{7}$ **(c)** $\frac{5}{6} + \frac{7}{12}$ **(d)** $\frac{2}{3} + \frac{5}{6}$
  (e) $1\frac{2}{5} + \frac{3}{15}$ **(f)** $2\frac{5}{6} + 1\frac{1}{3}$ **(g)** $3\frac{3}{4} + 1\frac{5}{8}$ **(h)** $\frac{9}{16} + \frac{3}{4}$

2 Find the sum of $2\frac{7}{16}$, $3\frac{1}{8}$ and $1\frac{3}{4}$.

3 Work out

  (a) $\frac{5}{6} - \frac{2}{6}$ **(b)** $\frac{7}{12} - \frac{1}{6}$ **(c)** $\frac{3}{4} - \frac{1}{8}$ **(d)** $2\frac{2}{5} - 1\frac{1}{10}$
  (e) $\frac{7}{16} - \frac{2}{8}$ **(f)** $3\frac{7}{8} - 1\frac{3}{4}$ **(g)** $2\frac{1}{5} - \frac{7}{10}$ **(h)** $3\frac{2}{9} - \frac{11}{27}$
  (i) $11\frac{4}{15} - 8\frac{3}{5}$ **(j)** $4\frac{4}{25} - 1\frac{3}{5}$

> **Remember:** To add or subtract fractions you need a common denominator.
> Add or subtract the numerators only.

4 Work out

(a) $\frac{2}{7} + \frac{1}{6}$ (b) $\frac{3}{5} + \frac{3}{8}$ (c) $\frac{5}{9} - \frac{2}{7}$ (d) $\frac{3}{10} - \frac{1}{11}$

(e) $2\frac{3}{4} + 3\frac{1}{5}$ (f) $5\frac{1}{6} + 3\frac{2}{9}$ (g) $3\frac{6}{15} - 2\frac{7}{10}$ (h) $1\frac{9}{16} - \frac{5}{24}$

Exercise 4.5   Links 4K–M

1 Work out the following. Give your answers in the simplest form.

(a) $\frac{1}{4} \times \frac{3}{8}$ (b) $\frac{2}{7} \times \frac{1}{3}$ (c) $\frac{5}{18} \times \frac{2}{15}$ (d) $\frac{3}{5} \times \frac{15}{24}$

(e) $1\frac{2}{9} \times \frac{3}{5}$ (f) $2\frac{1}{4} \times 1\frac{5}{18}$ (g) $3\frac{7}{10} \times 1\frac{4}{11}$ (h) $3\frac{5}{7} \times 2\frac{2}{13}$

2 Work out the following. Give your answer as a fraction in its simplest form.

Remember: Write mixed numbers as improper fractions.

(a) $\frac{3}{5} \div 1\frac{2}{3}$ (b) $1\frac{2}{7} \div \frac{3}{7}$ (c) $2\frac{4}{5} \div 2\frac{1}{3}$ (d) $3\frac{1}{5} \div \frac{7}{8}$

(e) $8\frac{1}{2} \div \frac{3}{8}$ (f) $4\frac{1}{5} \div 1\frac{1}{6}$ (g) $1\frac{2}{3} \div \frac{5}{18}$ (h) $2\frac{2}{7} \div \frac{4}{21}$

(i) $\frac{32}{343} \div 1\frac{15}{49}$ (j) $2\frac{27}{49} \div 3\frac{4}{7}$

3 Work out

(a) $\frac{2}{5}$ of 35 cm   (b) $\frac{3}{7}$ of 210 kg   (c) $\frac{3}{8}$ of £12

(d) $\frac{5}{6}$ of 48   (e) $\frac{5}{12}$ of £2160   (f) $\frac{2}{3}$ of 111

4 A bicycle journey of 56 miles is partly on roads and partly on tracks. $\frac{2}{7}$ of the journey is on tracks. How far is this?

5 Write 55p as a fraction of £3. Give your answer in its simplest form.

6 In a sports club $\frac{3}{5}$ of the members are men over 18 and $\frac{2}{7}$ are women over 18. What fraction are aged 18 and under?

7 A wall measuring 1.8 metres by 3.6 metres is tiled with 15 cm square tiles. $\frac{2}{9}$ of the tiles are pattern tiles. How many pattern tiles are used?

How many rows of tiles are there?
How many in each row?

8 A chess grandmaster is allowed $2\frac{1}{4}$ minutes for each move in a game. On average, he only takes $\frac{7}{8}$ of the time allowed. How much time does he use in a game taking 56 moves?

He makes 28 moves.

5 Collecting and recording data

Exercise 5.1 Link 5A

1   For each of these sets of data write down whether it is qualitative or quantitative.

(a)   the height of students at Lucea School

(b)   the students' hat size

(c)   the students' favourite food

(d)   the students' colour of eyes

> **Remember:**
> Qualitative data is described using words. Quantitative data is counted or measured.

2   For each of these types of data write down whether it is discrete or continuous.

(a)   the length of a TV programme

(b)   the number of apples in a bag

(c)   the number of cars in a car park

(d)   the weight of a sack of potatoes

(e)   the number of points scored by a rugby team

> **Remember:**
> Quantitative data can be discrete or continuous. Discrete data can take particular values. Continuous data can take any value.

3   For question **1** write down whether the data can be described as either discrete or continuous.

4   For question **2** write down whether the data is qualitative or quantitative.

5   For each type of data write down whether it is qualitative, quantitative and discrete, or quantitative and continuous.

(a)   the speed at which a horse can gallop

(b)   the marks in a test

(c)   the height of a mountain

(d)   the result of a game

(e)   the feel of a surface

(f)   the volume of a sphere

(g)   the thickness of custard

(h)   the capacity of a box

(i)   the age, to the nearest year, of a building

Exercise 5.2 Links 5B, C

1   There are 1800 students at Shimpling High School. The table shows how these students are distributed by year group and gender.

Year group	Number of boys	Number of girls
9	180	195
10	189	218
11	169	191
12	162	184
13	150	162

Philip is conducting a survey about the students' favourite television programmes. He decides to use a stratified random sample of 250 students according to year group and gender.

(a)   How many Year 12 boys should be in his sample?

(b)   How many Year 10 girls should be in his sample?

2   Explain how to make a selective sample of 8% of the 1500 workers at a car factory.

> **Remember:**
> A systematic (or selective) sample is one in which every nth item is chosen.

3   There are 1800 students at Irving Academy. One student, Sharon, wishes to take a random sample of 90 students for her maths project. Describe at least three different ways Sharon could take such a sample.

4   Here are some questions that do not work properly.
Improve each one and say what was wrong with the original.

(a)   Terry wants to carry out a survey on the number of hours that people watch television. He asks:
'Do you watch television?

a lot ☐   a little ☐   never ☐'

(b)   Andy is carrying out a survey on crime in his community.
He asks:
'Are the police OK?'

(c)   Anthea is carrying out a survey on favourite breakfast cereals.
She asks:
'Do you like cereals?'

(d)   Stina wants to find out what people think about paying for dental care. He asks:
'Do you agree that all dental care should be free?'

(e)   Michael wants to find out what type of music people like to listen to. He asks:
'Do you like pop music or classical music?'

(f)   Sethina wants to find out about the types of things people recycle. She asks:
'What do you recycle?

Paper ☐   Glass ☐'

> **Remember:** When you are writing questions for a questionnaire:
> - be clear about what you want to find out, and what data you need
> - ask simple questions
> - avoid questions which are too personal, or which are biased.

5 Design a questionnaire to investigate whether the council should provide more leisure facilities.

6 Design a questionnaire to investigate whether hamburgers are more popular with people under 16 or over 16.

Exercise 5.3 **Links 5D, E**

1 In each of these cases choose which type of sampling technique you would use to collect your data.

   (a) Where people eat their lunch
      (i) by asking the first 10 people in the lunch queue
      (ii) by asking every third person in your tutor group
      (iii) by asking the first 10 people going home for lunch.

   (b) What football team do you support
      (i) by asking people at a football match
      (ii) by asking people at a rugby match
      (iii) by asking a random sample of people in a telephone directory.

   (c) How you travelled to school
      (i) by asking every third person in your tutor group
      (ii) by asking all those people who were late for school
      (iii) by asking all those people on the school bus.

2 A new supermarket is to be built. The company carry out a survey, to find people's views on where it should be built. There are three sites where it could be built. One site is in the middle of town near three other supermarkets, one site is on the edge of town with its own free car park and the other site is next to the football ground.

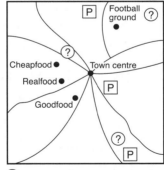

(?) Proposed new supermarket

   (a) Suggest a suitable sample and where it should be collected from.

   (b) Design a suitable questionnaire that the company could use to collect their data.

3 Would you use primary or secondary data for the following projects?

   (a) research into UK Olympic medals

   (b) the depth of polar ice caps

   (c) the country of origin of vegetables in your local supermarket.

> Data collection sheet is another name for data capture sheet.

4 Design a suitable data collection sheet to find out about the numbers and types of coins that people are carrying with them. Use your data collection sheet to check how easy it is to use.

Exercise 5.4   **Links 5F–H**

1 Robin had a holiday job packing cheese. Each pack of cheese should weigh 500 grams. Robin had to pack 30 packs of cheese. Robin checked the weights, in grams, correct to the nearest gram. These are the results:

512	506	503	506	499	499	500	504	502	503
496	497	497	509	506	499	497	498	507	511
498	491	496	506	507	493	496	503	510	508

Copy and complete the grouped frequency table for the weights. Use class intervals of 5 grams.

Weight (w grams)	Tally	Frequency
$490 \leqslant w < 495$		

[E]

2 The weights of 30 boys, in kilograms, in a class are given below.

50.7	55.1	48.3	52.3	50.9
61.3	54.7	59.8	53.2	46.5
63.7	53.0	56.9	54.2	46.9
52.8	66.3	54.1	51.3	54.5
57.4	52.6	48.2	64.2	57.1
55.9	60.3	54.2	41.1	54.3

(a) Draw up a table to display this data.

(b) Display the data as a histogram.

> **Remember:** A histogram looks like a bar chart but:
> - the data is continuous, so there can be no gaps between the bars
> - the data must be grouped into class intervals of equal width if you want to use the lengths of the bars to represent frequencies.

3 The table below contains some of the information about attendance at a rock festival.

	Male	Female	Total
20 and under		14%	24%
21–24	23%	21%	
25 and over			
Totals	57%		

Copy and complete the table.

(a) What was the percentage of females at 25 and over?

(b) What was the percentage of males at 25 and over?

4 The table shows some data about supermarket coffee sales.

	100 g	200 g	400 g	Total
Store brand A		126		
Brand B granules	120		53	
Brand B powder	41	86		147
Brand C powder	66		31	193
Total	300		132	

The manager remembers that they did not have any stocks of Brand B 200 g granules to sell. That enables the table to be completed.

(a) What was the total number of jars of coffee sold?

(b) How many 200 g jars of coffee were sold?

5 50 pupils are going on an educational visit. The pupils have to choose to go to one of the theatre, the art gallery and the science museum.
23 of the pupils are boys.
11 of the girls choose to visit the theatre.
9 of the girls choose to visit the art gallery.
13 of the boys choose to visit the science museum.
6 more pupils visit the art gallery than visit the theatre.
Draw up and complete a two-way table.
How many of the girls choose to visit the science museum?   [E]

6 Data on the distance and bearing (but not the altitude) of aircraft in the vicinity of Heathrow is taken at noon one day in the peak summer season. This data is given below in coordinate form (distance, bearing).

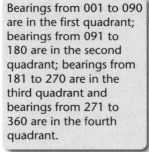

Bearings from 001 to 090 are in the first quadrant; bearings from 091 to 180 are in the second quadrant; bearings from 181 to 270 are in the third quadrant and bearings from 271 to 360 are in the fourth quadrant.

(5, 123)	(7, 080)	(14, 130)	(9, 275)	(11, 189)	(3, 021)
(2, 030)	(1, 039)	(6, 147)	(8, 152)	(14, 190)	(17, 005)
(23, 320)	(11, 216)	(12, 234)	(30, 300)	(18, 010)	(6, 095)
(4, 130)	(8, 235)	(11, 175)	(6, 125)	(9, 300)	(16, 165)
(12, 060)	(14, 130)	(1, 220)	(7, 215)	(11, 155)	(29, 081)
(13, 110)	(17, 215)	(14, 280)	(4, 290)		

Construct and complete a two-way table with the distance categories $0 < x < 5$, $5 \leqslant x < 10$, $10 \leqslant x < 15$, $x \geqslant 15$ and the bearing categories as the four quadrants.
Which is the most crowded region?

7 The frequency distribution gives the weekly sales of dresses in Bettie's Boutique for one year.
Draw up a frequency polygon for this data.

Remember: For a frequency polygon, plot the frequencies against the midpoints of the class intervals.
The midpoint of the interval 20–29 is
$\dfrac{20 + 29}{2} = 24.5$, so plot (24.5, 1).

Number of dresses sold	20–29	30–39	40–49	50–59	60–69	70–79	80–89
Number of weeks	1	4	19	13	8	5	2

8 This frequency polygon shows the times that 30 boys take to do a puzzle.

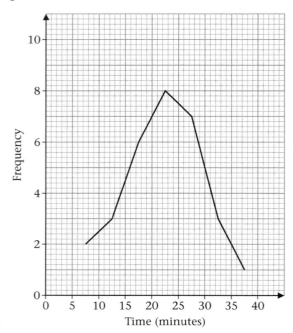

A group of 30 girls were asked to do the same puzzle. Here are their times, each to the nearest minute.

12	24	16	5	12	17
22	32	23	21	21	14
18	16	9	16	11	26
8	27	24	26	17	17
23	19	28	34	29	19

(a) Group the data and draw a frequency polygon.

(b) Compare the two frequency polygons. What conclusions can you draw from the two frequency polygons?

6 Solving equations and inequalities

Exercise 6.1

Links 6A, B

Solve the equations and check your answers by substitution.

1 $47 = 5m + 2$ **2** $48 = 4n + 20$ **3** $37 = 5p - 3$

4 $40 = 4(x + 1)$ **5** $8 = 2(y - 3)$ **6** $3 = \dfrac{b}{4} + 8$

7 $3h - 5 = 7$ **8** $5x + 2 = 22$ **9** $6x - 1 = 11$

10 $7 + x = 9$ **11** $\dfrac{m}{4} - 1 = 4$ **12** $\frac{1}{3}x + 2 = 7$

13 $\frac{1}{4}x + 5 = 16$ **14** $\frac{1}{5}x - 3 = 2$ **15** $4 - \frac{1}{3}x = 6$

16 $\frac{2}{3}x + 9 = 4$ **17** $5 - \frac{3}{4}x = 17$

Exercise 6.2

Links 6C, D

1 Solve each of these equations.

(a) $3x + 7 = 22 - 2x$ (b) $6x - 3 = 13 - 2x$

(c) $5y - 4 = 26 - 5y$ (d) $1\frac{1}{2}x + 2 = 8 - \frac{1}{2}x$

(e) $4a + 1 = 8 - 3a$ (f) $8n + 11 = 1 + 6n$

(g) $3 - 4x = 12 - 7x$ (h) $y + 1 = 4 - 2y$

(i) $17 + m = 31 - 6m$ (j) $6 - x = 11 - 3x$

(k) $5 - 4x = 3 + x$ (l) $7 + 3y = 5y - 2$

Remember: Collect the unknown on one side and numbers on the other.

2 (a) $3x + 1 = 2x + 4$ (b) $5x - 7 = 4x + 4$

(c) $16x + 1 = 12x + 1$ (d) $2 - 3x = 2x + 7$

(e) $19x + 13 = 17x + 21$ (f) $15k - 3 = 4 + 6k$

(g) $13 - 3p = 4 - 2p$ (h) $8 - 5q = 21 - 2q$

(i) $9 + 3q = 2q - 6$

Exercise 6.3

Links 6E, F

1 Solve each of these equations.

(a) $3(x + 1) = 15$ (b) $2(y + 3) = y + 11$

(c) $4(n - 3) = 3(n + 1)$ (d) $5(x + 2) = 3x + 16$

(e) $7x - 3 = 3(2x + 1)$ (f) $4y = \frac{1}{2}(2y + 3)$

(g) $5 + 3x = 2(2x + 3)$ (h) $3(1 - 2y) = 8$

(i) $5(3n + 1) = 12n + 14$ (j) $3(4 - 5y) = 4(6 - 5y)$

(k) $7(9 - y) = 5(3y - 5)$ (l) $4(5 + x) = 5(x - 3) + 22$

(m) $\dfrac{a + 3}{4} = 2$ (n) $\dfrac{2b - 3}{5} = 7$

(o) $\dfrac{3n - 6}{4} = 1\frac{1}{2}$ (p) $\dfrac{4a + 2}{6} = 7$

It is not always best to expand the brackets. Deal with fractions first.

2 Solve each of these equations.

Multiply both sides by the lowest common multiple (LCM).

(a) $\dfrac{x}{2} + 3 = \dfrac{x}{4} + 5$

(b) $\dfrac{y}{3} + 3 = \dfrac{y}{4} + 4$

(c) $2x - 1 = \dfrac{3x}{2} + 6$

(d) $\dfrac{n + 5}{3} = \dfrac{n}{7} + 3$

(e) $\dfrac{x}{3} = 5(x - 2)$

(f) $\dfrac{3a}{4} - 1 = 5(a - 1)$

(g) $\dfrac{b + 1}{5} = \dfrac{b - 1}{4}$

(h) $5(3x - 2) = \dfrac{2x + 3}{2}$

3 Solve each of these equations.

(a) $7 - 4x = 15$ (b) $3(2y - 1) = y$ (c) $4(n - 3) = \dfrac{n}{5}$

Exercise 6.4 **Link 6G**

1 The sum of three consecutive numbers is 45. Find the numbers.

Let the first number be x. Write the problem as an equation.

2 In two years' time I will be 3 times as old as I am now. How old am I?

3 If I multiply the number I am thinking of by 5 and subtract 3, the answer is 72. What number am I thinking of?

4 The perimeter of this quadrilateral is 55. Find x.

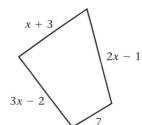

5 The diagram shows two of the sides of a regular pentagon.

(a) Write down an equation in x which must be true.

(b) Solve this equation.

(c) Work out the perimeter of the pentagon.

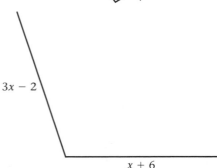

6 At the start of the week, Alex has £20 and Mary has £17. During the week Alex hires four videos and Mary hires two videos. Each video costs the same amount to hire. Neither of them spends any other money. At the end of the week they both have the same amount of money left. Let x stand for the cost, in £, of hiring a video.

(a) Write down an equation in x which must be true.

(b) Solve the equation to find the cost of hiring a video.

7 The triangle and the pentagon both have the same perimeter.
Write down an equation and solve it to find the value of x.

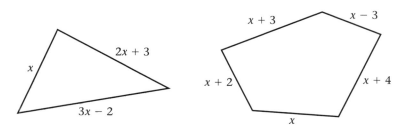

Exercise 6.5

Links 6H–J

1 Write down the inequalities shown on the number lines.

(a)

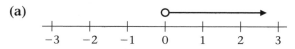

(b)

(c)

(d)

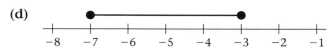

(e)

> **Remember:**
> ● includes the number
> ○ excludes the number

2 Draw number lines to show the inequalities.

   (a) $x > 2$ **(b)** $x \leqslant 3$ **(c)** $x \geqslant -2$

   (d) $-3 < x < 3$ **(e)** $-1 \leqslant x < 0$ **(f)** $-2 < x \leqslant 3$

> **Remember:**
> \> greater than
> \< less than
> ⩾ greater than or equal to
> ⩽ less than or equal to

3 Solve these inequalities.

   (a) $2x - 1 \leqslant 5$ **(b)** $3x + 2 > 14$ **(c)** $4x + 1 \geqslant 6$

   (d) $2x + 13 \leqslant 2$ **(e)** $3 - 2x < 1$ **(f)** $14 - 3x > 8$

   (g) $3s + 1 < 6 - 2s$ **(h)** $4 + 3s \geqslant s - 7$ **(i)** $6t + 1 \leqslant 2t + 1$

   (j) $3 - 5v > 2 - 2v$ **(k)** $17 - 3v \leqslant 5v - 3$

4 Solve these inequalities.

   (a) $3 < 2k + 1 < 21$ **(b)** $4 < 2 + 2q \leqslant 5$

   (c) $-3 \leqslant 5r + 2 < 17$ **(d)** $-1 < 2 + \frac{1}{2}s < 7$

Exercise 6.6 Links 6K, L

1 List the possible integer values that satisfy these inequalities.

(a) $-3 < x < 1$ (b) $-1 \leqslant x < 4$

(c) $-6 < x \leqslant -3$ (d) $-2 < 3x + 1 \leqslant 7$

(e) $-5 \leqslant 1 - 4x \leqslant 18$

2 Write down an inequality satisfied by these integers.

(a) $0, 1, 2$ (b) $2, 3, 4, 5, 6$ (c) $-8, -7, -6, -5$

(d) $-1, 0, 1$ (e) $22, 23$ (f) $-3, -2, -1$

3 Find the greatest integer value of n.

(a) $5n + 3 < 14$ (b) $6 - 3n \geqslant 2$ (c) $7n + 4 < 1$

(d) $4 - 2n > 7$ (e) $3 + 3n < 10 - n$ (f) $3 - 2n \geqslant 6 - n$

> **Remember:** You must not multiply or divide an inequality by a negative number.

4 Find the smallest integer value of n.

(a) $2n + 7 \geqslant 1$ (b) $9 - 3n < 6$ (c) $5n + 9 > 0$

(d) $8 - 5n \leqslant -6$ (e) $0 > 3 - 2n$ (f) $5n - 5 > n + 20$

5 x is an integer, such that $-3 < x \leqslant 2$.
List all the possible values of x.   [E]

6 List all the possible integer values of x.

(a) $5 < 1 + x < 8$ (b) $-7 < 2x + 3 \leqslant 3$

(c) $0 \leqslant 4 - 2x < 15$ (d) $-8 < 2 - 3x < 1$

> **Remember:** Treat a two-sided inequality as two inequalities and combine the solutions.

7 Transformations and loci

1 Copy this grid into your exercise book.
Translate the shape T using the following vectors.

(a) $\binom{5}{2}$; call it A.

(b) $\binom{5}{-2}$; call it B.

(c) $\binom{-4}{-2}$; call it C.

(d) $\binom{-1}{1}$; call it D.

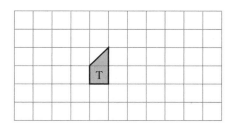

Remember:

In a vector $\binom{a}{b}$

a is the movement to the right
b is the movement upwards.

2 The shape S has been moved into four different positions A, B, C, D.
Write down the vector for each of the four translations.

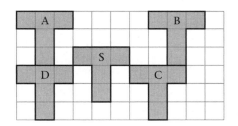

3 Copy this diagram into your exercise book.
For each shape reflect it in the mirror line **m**.

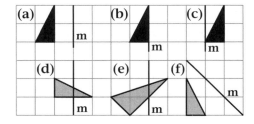

4 Shape S has been reflected four times.

(a) Describe fully each of these reflections, which take S to positions A, B, C, D.

(b) Write down the rotation that moves shape A on to shape C.

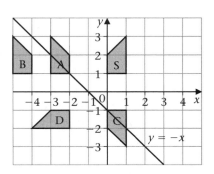

Remember: To describe a reflection you need to give the equation of the mirror line.
A rotation is described by giving
● the centre of rotation
● the amount of turn
● the direction of turn.

5 Copy this diagram into your exercise book.
Rotate each triangle about the point P marked
with a cross by the angle and direction given.

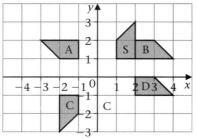

(a)   +90°   (b)   −270°

(c)   −180°   (d)   −90°

(e)   +180°   (f)   +45°

6 Shape S has been rotated four times.

(a)   Describe fully each of these rotations, which take S to
positions A, B, C, D.

(b)   Write down the rotation that moves shape C on to
shape A.

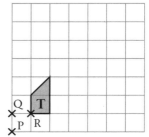

7 Copy this diagram into your exercise book.
Enlarge shape T by

(a)   scale factor 2 from point P

(b)   scale factor 3 from point Q

(c)   scale factor 2.5 from point R

(d)   scale factor $\frac{1}{2}$ from point R

(e)   scale factor $-\frac{1}{2}$ from point R

> **Remember:** An
> enlargement is
> described by giving
> • its centre of
>   enlargement
> • its scale factor.

8 Describe the enlargement that moves

(a)   shape P on to shape Q

(b)   shape T on to shape R

(c)   shape T on to shape S

(d)   shape R on to shape T

(e)   shape T on to shape V

(f)   shape P on to shape W

(g)   shape W on to shape P

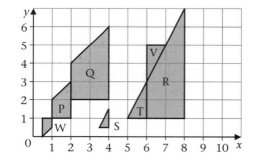

Exercise 7.2   Link 7E

1 (a)   Triangle T is transformed on
to triangle R by a single
transformation. Write down
in full this single transformation.

(b)   Triangle R is transformed on to
triangle M by a single
transformation. Write down
in full this single transformation.

(c)   Write down in full the single
transformation that will move triangle M back on to triangle T.

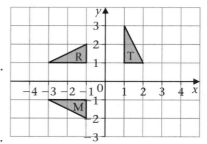

> A single transformation
> is just one
> transformation, not
> one followed by
> another.

2 On a coordinate grid with both x- and y-axes drawn from -4 to $+4$ plot the points (1, 1), (1, 3), (2, 2) and (2, 1). Label this shape A.

   (a) Transform shape A by reflecting it in the x-axis. Label the shape B.

   (b) Rotate shape B by $-90°$ with centre (0, 0). Label this shape C.

   (c) Write down the single transformation that will move shape C back on to shape A.

3 On a coordinate grid with both x- and y-axes drawn from -4 to $+4$ plot the points (1, 1), (1, 3), (2, 2) and (2, 1). Label this shape A.

   (a) Transform shape A by reflecting it in the y-axis. Label the shape B.

   (b) Rotate shape B by $90°$ with centre (0, 0). Label this shape C.

   (c) Write down the single transformation that will move shape C back on to shape A.

4 On a coordinate grid with both x- and y-axes drawn from -6 to $+6$ plot the points (1, 1), (1, 3), (2, 2) and (2, 1). Label this shape A.

   (a) Transform shape A by enlarging it by scale factor 2 centre (0, 0). Label the shape B.

   (b) Rotate shape B by $90°$ with centre (0, 0). Label this shape C.

   (c) Reflect shape C by reflecting it in the y-axis. Label this shape D.

   (d) Write down the single transformation that will move shape D back on to shape B.

5 On a coordinate grid with both x- and y-axes drawn from -4 to $+4$ plot the points (1, 1), (1, 3), (2, 2) and (2, 1). Label this shape A.

   (a) Rotate shape A by $90°$ with centre (0, 0). Label this shape B.

   (b) Transform shape B by reflecting it in the x-axis. Label the shape C.

   (c) Reflect shape C in the line $y = x$. Label the reflected shape D.

   (d) Transform shape D by rotating it by $180°$ centre $(-1, -1)$. Label the new shape E.

   (e) Write down the single transformation that will move shape E back on to shape A.

6 Copy the diagram.

(a) Reflect the triangle T in the line $y = x$.
Label the shape R.

(b) Rotate the triangle T through 180° about (0, 0).
Label the shape S.

(c) Describe fully the transformation that will move
shape S on to shape R.

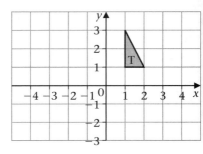

Exercise 7.3   Link 7F

1 Albert makes a model of a fire engine. The scale of the model is
1 : 20. The height of the real fire engine is 4.2 metres.

(a) Calculate the height of the model fire engine.

The length of the model fire engine is 54 cm.

(b) Calculate the length of the real fire engine.

The ladder makes an angle of 15° with the top of the model fire
engine.

(c) What angle does the ladder make with the top of the real fire
engine?

> To make it bigger,
> multiply by 20. To make
> it smaller, divide by 20.

2 On the Pathfinder map she was using, Gill measured the distance
she still had to walk as 14.5 cm. The map was drawn on a scale of
1 : 25 000. Calculate the actual distance Gill still had to walk.

3 Caesar was driving along Ermin Street and measured the distance
he travelled in a straight line as 12 kilometres. What distance
would that be on a map with a scale of 1 : 50 000?

4 Sue submitted plans for an extension to her house. The scale of
the plans was 1 : 50. The length of the extension on the plan was
5.3 cm.

(a) What was the actual length of the extension?

The width of the actual extension is 3.1 metres.

(b) What measurement would the width be on the plan?

Exercise 7.4   Link 7G

1 Using only a ruler, compasses and pencil, construct triangles

(a) *ABC* so that *AB* = 12 cm, *BC* = 8 cm, *AC* = 6 cm

(b) *PQR* so that *PQ* = 6 cm, *QR* = 6 cm, angle *Q* = 60°

(c) *XYZ* so that *XY* = 8 cm, *YZ* = 6 cm, angle *Y* = 30°

> **Remember:** Show all
> your construction lines.

> 30° = half of 60°, so
> bisect 60°.

2 Draw a line *AB* 8.4 cm long. Draw the perpendicular bisector of *AB*.

> Put your compass point on *A* and *B* in turn.

3 Draw a triangle *ABC* so that *AB* = 8 cm, *BC* = 8 cm, *AC* = 10 cm.

   (a) Draw the perpendicular bisectors of *AB*, *BC* and *AC*. The three perpendicular bisectors should meet at one point.

   (b) Place your compass point at this point and draw a circle that passes through points *A*, *B* and *C*. This is called the circumscribed circle of triangle *ABC*.

4 Draw a triangle *PQR* so that *PQ* = 8.5 cm, *QR* = 6.5 cm, *PR* = 9.5 cm.

   (a) Draw the angle bisectors of ∠*P*, ∠*Q* and ∠*R*. The three angle bisectors should meet at one point.

   (b) Place your compass point at this point and draw a circle that touches *PQ*, *QR* and *PR*. This is called the inscribed circle of triangle *PQR*.

Exercise 7.5

Link 7H

1 Construct the locus of the following points.

   (a) 3 cm from the point *Q*

   (b) equidistant from *X* and *Y* where *XY* = 5 cm

   (c) equidistant from the lines *AB* and *BC* where angle *ABC* = 60°

   (d) 2 cm from the straight line *PQ* where *PQ* = 7.5 cm

2 A goat is tethered by a 10-metre-long chain in the middle of a large field. Draw, using a scale of 1 cm to represent 4 metres, the locus of the area that the goat can graze in if the chain is attached

> **Remember:** Show all your construction lines.

   (a) to a tree   **(b)** to a bar that is 20 metres long.

3 Ermintrude the cow is attached by a 15-metre-long chain to a bar that runs along the long side of a barn that is located in the middle of a large field as shown in the diagram. Using a scale of 1 cm to represent 2 metres draw the locus of the area in which she can graze.

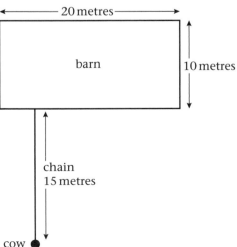

Exercise 7.6   **Link 7I**

1 Write down the three-figure bearing for each of these directions.

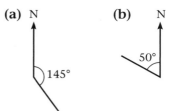

(a) N   145°

(b) N   50°

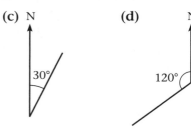

(c) N   30°

(d) N   120°

2 Write down the three-figure bearing for each of these directions.

(a) south	**(b)** west	**(c)** east
(d) south-west	**(e)** north-east	**(f)** north-west
(g) north	**(h)** south-east	

3 Measure and write down the bearing of *B* from *A* in each of these diagrams.

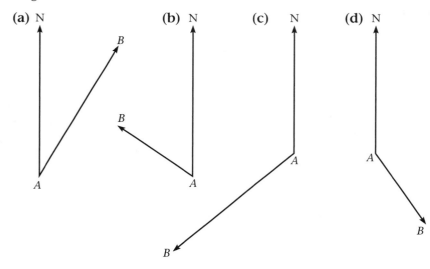

(a) N   **(b)** N   **(c)** N   **(d)** N

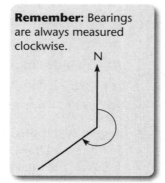

Remember: Bearings are always measured clockwise.

4 For each of the bearings in question **3**, work out the bearing of *A* from *B*.

5 Draw diagrams to show the following bearings.

(a) *Y* on a bearing of 060° from *X*

(b) *P* on a bearing of 145° from *Q*

(c) *R* on a bearing of 310° from *S*

(d) *G* on a bearing of 220° from *H*

8 Using basic number skills

1 Work out

(a) 10% of £15

(b) 12% of £18

(c) $17\frac{1}{2}$% of £50

(d) $17\frac{1}{2}$% of £185

(e) 5% of 30 metres

(f) 24% of 45 centimetres

(g) $37\frac{1}{2}$% of 150

(h) $66\frac{2}{3}$% of 297

> To find 15% multiply by 0.15

2 Express the first quantity as a percentage of the second.

(a) £35, £50

(b) £45, £60

(c) 20p, 30p

(d) 24 m, 40 m

(e) £17, £20

(f) £12, £8

(g) 5 m, 6 m

(h) 5 cm, 4 cm

> To express £7 as a percentage of £8 work out $\frac{7}{8} \times 100$

3 (a) Increase £50 by 5%

(b) Increase £120 by $17\frac{1}{2}$%

(c) Decrease £30 by 12%

(d) Decrease £135 by $12\frac{1}{2}$%

(e) Increase 56 cm by 8%

(f) Decrease 1.2 m by 15%

(g) Increase £3200 by 6%

(h) Decrease £2500 by $33\frac{1}{3}$%

> To increase by 15% multiply by 1.15
> To decrease by 15% multiply by 0.85

4 Gareth weighs 12 stone. He needs to lose 2 stone. What percentage of his weight is this?

5 Susan weighed 85 kg before going on a diet. She lost 6% of her original weight. What weight did she end up at?

6 Seamus buys a car for £14 000. He sells it a year later for £12 500. By what percentage has the car depreciated?

7 By what number do you multiply to

(a) increase a quantity by 12%

(b) decrease a quantity by 15%

(c) increase a quantity by 8%

(d) decrease a quantity by 6%

(e) add VAT at $17\frac{1}{2}$%

(f) give a sale price with 20% off?

8 Calculate the percentage increase or decrease in each case.

(a) £25 to £30

(b) £2 to £2.50

(c) $3000 to $5000

(d) 62 kg to 54 kg

(e) £56 to £64

(f) £84 to £72

> **Remember:**
> Percentage increase
> $= \dfrac{\text{actual increase}}{\text{original amount}} \times 100$

9 Rodney buys 100 mobile phones for £5000. He sells them all for £39.95 each. Work out his percentage loss.

10 Tricky Dicky bought 100 mobile phones for £5000. He sold them at £89.90 each. Work out his percentage profit.

11 In 2005 Roy bought a brand new car for £15 000. It is going to lose 15% of its value in the first year and 10% in every year after that. Work out the value of the car after 5 years.

Exercise 8.2   Links 8E, F

1 In 2000 Rashmi bought a house for £160 000. It lost 2% of its value during the first year and then gained 3% of its value over each of the next 2 years. At the end of the fourth year the house was worth £180 000. What was the percentage increase or decrease in value of Rashmi's house during the fourth year?

2 Work out the original price in the following cases.

	New price	Percentage change	Original price
(a)	£150	10% increase	
(b)	£150	10% decrease	
(c)	£5	30% decrease	
(d)	£3500	15% increase	
(e)	£2000	2% increase	
(f)	£19.50	5% decrease	
(g)	£13.25	5% increase	
(h)	£2.99	$17\frac{1}{2}$% increase	

Remember: To find the original price divide by the multiplier, e.g. to work back from an increase of 15% (×1.15) you divide by 1.15

3 Abdul buys a TV set in a sale. The TV set cost Abdul £350 after a discount of 20%. How much money did Abdul save by buying in the sale?

4 Cecile buys a Chinese silk carpet for £500 after VAT at $17\frac{1}{2}$% has been added. What was the price before VAT was added?

5 Simon bought a house for £150 000 in 2003. A year later it had increased in value by 10%. The year after, it decreased in value by 5%. What was the house worth in 2005?

6 Ruth bought a car for £5000 in 1960. In 1970 the car was worth 10% of its value in 1960. In 2000 the car was worth 50% more than it was worth in 1970. What was the car worth in 2000?

Exercise 8.3 Link 8G

1 Cheryl invests £200 at 4% compound interest for 3 years. How much money is in the account at the end of this time?

> Multiply by 1.04 for one year, but multiply by $(1.04)^4$ for four years, or $1.04 \times 1.04 \times 1.04 \times 1.04$

2 Work out the total amount of money when the following amounts of money are invested at compound interest.

   (a) £200 for 2 years at 5%

   (b) £1000 for 3 years at $4\frac{1}{2}$%

   (c) £250 for 4 years at 3.6%

   (d) £10 000 for 5 years at 6.1%

3 Piara invested £1000 at $5\frac{1}{2}$% compound interest for 5 years.

   (a) How much money is in the account at the end of 5 years?

   (b) What is the difference in the amount of interest if Piara had used simple interest and not compound interest?

4 Davindar invests money at 8.1% compound interest. How many years must the money be invested if it is to double in value?

5 Rachael borrows £5000 from the bank at an interest rate of 10%. She pays it back at the rate of £1000 per annum. Copy this table, extend it and use it to work out how many years it takes Rachael to repay the loan.

Year	Amount owed at start of year	Working	Amount owed at end of year
1	£5000	10% of £5000 = £500	£5000 + 500 − 1000 = £4500
2	£4500		

Exercise 8.4 Links 8H–K

1 Bill gets in his car to travel from London to Edinburgh. The distance is 400 miles and it takes Bill 8 hours to get there. What is the average speed for the whole journey?

> **Remember:**
> $$\text{Speed} = \frac{\text{distance}}{\text{time}}$$
> Distance = speed × time

2 Nik travels 80 miles by train in 1 hour 5 minutes. Work out the average speed in miles per hour.

3 Natalie travels 85 miles from her home to London. The first 5 miles take 15 minutes, the next 70 miles take 65 minutes and the last 10 miles take 55 minutes.

> **Remember:**
> $$\text{Average speed} = \frac{\text{total distance}}{\text{total time}}$$

   (a) Work out her average speed for each of the three stages of the journey.

   (b) Calculate the average speed for the whole journey.

4 Keith and David are taking part in a 20 mile sponsored walk. Keith takes 6 hours 20 minutes and David takes 40 minutes less. Work out their average speeds.

5 Daniel's car has a trip computer. It tells him how many miles per gallon of petrol he is getting. He is travelling from his home in the middle of Manchester to see his mother who lives in the middle of Bristol. The total distance is 165 miles. The table shows how many miles per gallon the car does at each stage of the journey. Work out the average number of miles per gallon the car does on the whole journey.

Miles travelled	Miles per gallon
10	30
150	40
5	5

6 Kate used teletext to find the exchange rate between euros, pounds sterling and dollars. She found that

> £1 was equivalent to €1.45
> £1 was equivalent to $1.72

Calculate the exchange rate between euros and dollars.

> This means €1.45 = $1.72
> What is €1 worth in dollars?

7 A cylindrical water tank full of water has a diameter of 0.6 metres and a height of 1.2 metres. It is emptying into a drain at a rate of 250 ml per minute.

(a) How long will it take to empty the whole tank?

Once the tank is empty water is poured in at a rate of 300 ml per minute whilst still emptying at 250 ml per minute.

(b) How long does it take the tank to completely fill again?

> **Remember:**
> Volume of cylinder = $\pi r^2 h$
> 1 ml = 1 cm³

8 A water main has a diameter of 30 centimetres. The water flow through the pipe is at a speed of 5 metres per second. Calculate the amount of water that passes through the pipe in one hour.

> First calculate the volume of water in a one second stretch.

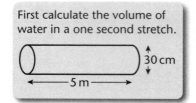

9 The density of gold is 19 355 kg/m³. Calculate the weight of a gold bar in the shape of a cuboid that has dimensions 20 cm by 8 cm by 25 mm.
The gold bar is recast into cylindrical gold wire without losing any of the metal. The cylinder has a diameter of 2 mm. How long is the wire?

10 There used to be 4 farthings in a penny, 4 pennies in a groat and 12 pennies in a shilling. Calculate the number of

(a) farthings in 25 groats

(b) groats in 20 shillings.

Exercise 8.5
Links 8L, M

1 Sandra shares a packet of 24 sweets with Mel in the ratio 5 : 3. How many sweets does each of them receive?

> Sandra has $\frac{5}{8}$ and Mel has $\frac{3}{8}$.

2 Richard and Ann share their household expenses in the ratio 3 : 2.

(a) Work out what each of them pays if the electricity bill is £35.

(b) Richard pays £12 as his share of the gas bill. What does Ann pay?

(c) Ann pays £12 as her share of the council tax. What does Richard pay?

3 Luke and Melissa give 10% of their take-home pay to charity. The ratio of their take-home pay is 5 : 4.

(a) Luke gives £20 to charity one week. How much does Melissa give?

(b) Melissa gives £20 to charity the next week. How much does Luke give?

(c) One week the total they gave to charity was £21.60. Work out their take-home pay that week.

4 The standard model car can travel 4.8 miles per litre of petrol. The deluxe model car can travel 4.5 miles per litre of petrol.

(a) Write down the ratio of their petrol consumption in a simpler form.

(b) Both cars travel 620 miles. How much petrol does each car use?

5 Peter, Paul and Mary buy lottery tickets in the ratio 2 : 3 : 4. They share their winnings in the same ratio. How much does each receive if they win

(a) £10   (b) £84   (c) £5000   (d) £2 000 000?

How much do you think we'll win?

Exercise 8.6
Links 8N–P

1 Write these numbers as powers of 10.

(a) 1000   (b) 100 000   (c) 10   (d) 1 000 000

(e) 1   (f) 10 000   (g) 100   (h) 100 000 000

2 Write these numbers in standard form.

(a) 3450   (b) 239   (c) 376 000   (d) 2 460 000

(e) 350   (f) 45 000   (g) 87   (h) 876 333

> **Remember:**
> $3521 = 3.521 \times 10^3$

3 Write these numbers, which are written in standard form, as ordinary numbers.

(a) 4.56×10^4 (b) 6.3×10^2 (c) 9×10^5

(d) 6.9×10^6 (e) 7.2×10^1 (f) 8.1×10^3

(g) 4.53×10^7 (h) 5.332×10^5

4 Write these numbers as powers of 10.

(a) 0.01   (b) 0.1   (c) 0.000 01

(d) 0.000 000 1   (e) 0.001   (f) 0.0001

(g) 0.000 001   (h) 0.000 000 01

5 Write these numbers, which are written in standard form, as ordinary numbers.

(a) 4.56×10^{-4} (b) 5.9×10^{-3} (c) 7×10^{-2}

(d) 3.75×10^{-6} (e) 5.6×10^{-1} (f) 7.25×10^{-7}

(g) 2.2×10^{-5} (h) 4.112×10^{-3}

> **Remember:**
> $0.035\,21 = 3.521 \times 10^{-2}$

6 Write these numbers in standard form.

(a) 0.0234   (b) 0.000 32   (c) 0.675

(d) 0.000 003 45   (e) 0.000 023   (f) 0.0893

(g) 0.002 895   (h) 0.000 000 561

Exercise 8.7   Link 8Q

1 Calculate these, giving your answers in standard form.

(a) $(4.54 \times 10^4) \times (3.76 \times 10^3)$

(b) $(2.225 \times 10^3) \div (3.65 \times 10^2)$

(c) $(5.6 \times 10^5) \times (6.95 \times 10^{-6})$

(d) $(7.25 \times 10^2) \div (9.75 \times 10^1)$

(e) $(5.76 \times 10^{-4}) \times (3.84 \times 10^{-3})$

(f) $(2.65 \times 10^{-2}) \div (6.34 \times 10^{-3})$

(g) $\dfrac{8.34 \times 10^3 + 5.98 \times 10^4}{6.53 \times 10^{-5}}$

(h) $\dfrac{6.98 \times 10^{-4} - 9.95 \times 10^{-5}}{5.75 \times 10^3}$

> Make sure you know how to enter numbers in standard form on your calculator.

2 The distance of the Sun from Mercury is 3.6×10^7 miles.
The distance of Neptune from the Sun is 2.79×10^9 miles.

(a) How many times further is it from the Sun to Neptune than to Mercury?

Light travels at 186 000 miles per second.

(b) How long does it take a ray of light to travel
(i) from the Sun to Mercury   (ii) from the Sun to Neptune?

3 The average distance between two atoms is 1.6 angstroms where 10^{10} angstroms are equal to 1 metre. Write 1.6 angstroms in metres in standard form.

4 A light year is the distance travelled by a ray of light in one year. The speed of light is approximately 3.0×10^5 kilometres per second. How far, in kilometres, is one light year? Give your answer in standard form.

9 Functions, lines, simultaneous equations and regions

Exercise 9.1   Link 9A

1 Work out the outputs from each function using the inputs
$n = 1, 2, 3, 10$

> Remember the order of operations.

(a) $n \to 3n - 2$ (b) $n \to n + 5$ (c) $n \to 2n + 7$

(d) $n \to 2(n + 5)$ (e) $n \to \dfrac{n}{2} + 5$ (f) $n \to 2(n - 1)$

(g) $n \to 2n - 2$ (h) $n \to \dfrac{n + 5}{2}$

2 Work out the outputs from each of the functions for the given inputs.

(a) $n \to n + 3$ for $n = 0, 2, 5, 17$

(b) $n \to 3n + 1$ for $n = 1, 3, 5, 7$

(c) $n \to 2n - 3$ for $n = 0, 1, 2, 3, 4$

(d) $n \to 3n + 2$ for $n = -2, -1, 0, 1, 2$

(e) $n \to 2(n + 3)$ for $n = 0, 1, 2, 3$

(f) $n \to 2n + 6$ for $n = 0, 1, 2, 3$

What do you notice about the outputs for the functions in (e) and (f)?

Exercise 9.2   Links 9B, C

1 (a) Write each of these functions as an equation using x and y.
   (i) $n \to n + 3$ (ii) $n \to n - 1$
   (iii) $n \to 3n$ (iv) $n \to 3n + 4$

(b) Make a table of values for each equation using the values
   0, 1, 2, 3, 4, 5 for x.

(c) Draw the graphs to show the numbers for each of your tables
   of values.

2 **(a)** Copy and complete the table of values for this sequence of dot patterns.

Stage x	1	2	3	4	5
Number of dots y	5	8	11		

(b) Draw the graph for this table of values.

(c) Write down the equation of this graph.

3 **(a)** Invent a sequence of matchstick patterns for the table of values below.

x	1	2	3	4	5
y	6	11	16	21	26

(b) Draw the graph for this table of values.

(c) Find the equation of this graph.

4 Two common Imperial units of measure are yards and feet. They are connected by the rule

$$1 \text{ yard} = 3 \text{ feet}$$

(a) Copy and complete this table of values.

Yards x	0	1	2	3	4	5	6
Feet y	0	3				15	

(b) Draw the graph for this table of values.

(c) Write down the equation of this graph.

(d) By using your graph, or otherwise, find
  (i) the number of feet equal to 2.5 yards
  (ii) the number of yards equal to 10 feet.

Exercise 9.3   **Links 9D, E**

1 On separate diagrams draw lines with these equations.
  (a) $y = x + 1$ **(b)** $y = 2x + 2$ **(c)** $y = 3x - 1$
  (d) $y = 4 - x$ **(e)** $y = 3 - 2x$ **(f)** $y = \frac{1}{4}x$

2 For each line in question **1**, write the equation of two lines which are parallel.

> **Remember:** Lines that are parallel have the same gradient (m).

3 Write down the intercepts on the y-axis of the lines with these equations.
  (a) $y = 2x + 3$ **(b)** $y = 4 - 2x$ **(c)** $y = 3x + 7$
  (d) $y = 2 + \frac{1}{2}x$ **(e)** $y = -\frac{1}{2}x - 5$ **(f)** $y = 3\frac{1}{2}x - 1$

4 A line parallel to $y = 2x + 3$ meets the y-axis at $(0, -2)$.
Write down the equation of the line.

5 A line with equation $y = 3x + a$ crosses the y-axis at $(0, 3)$.
Write the equation of the line.

6 A line with equation $y = c - 2x$ passes through the point $(2, 3)$.
Write down the equation of the line.

7 A line with equation $y = ax + 3$ passes through the point $(3, 0)$.
Work out the equation of the line.

8 Find the intercept on the x-axis of the lines
   (a) $y = 2x - 4$ (b) $y = 6 - 3x$
   (c) $y = \frac{1}{2}x + 7$ (d) $y = 14 - \frac{1}{3}x$

9 Write down the equation of the line with gradient 4 and
   y-intercept at $(0, 3)$.

10 Write down the equation of the line with gradient $-\frac{1}{2}$ and
   y-intercept at $(0, -4)$.

11 Write down the gradient and y-intercept of the lines
   (a) $2x + y = 7$ (b) $y - 3x + 2 = 0$
   (c) $2x + 3y = 5$ (d) $2x - 3y + 1 = 0$
   (e) $14 - 2x - 7y = 0$ (f) $-7 = 3x - 2y$

> Rearrange the equations into the form $y = mx + c$.

12 The gradient of a line is 5. It passes through the point $(1, 2)$.
Work out the equation of the line.

13 The point $(3, -2)$ lies on a line with gradient -3. Work out the
equation of the line.

14 Find the equation of the line that passes through the points $(2, 7)$
and $(14, 31)$.

15 Find the equation of the line which passes through the points
$(-3, -1)$ and $(0, 11)$.

16 Find the equation of the line which passes through the points
$(-3, 6)$ and $(3, 3)$.

Exercise 9.4
Links 9F, G

1 Pick out the pairs of perpendicular lines from this list of
equations.
   (a) $y = 2x + 7$ (b) $y = \frac{1}{2}x - 3$
   (c) $y = x + 4$ (d) $y = 3 - \frac{1}{2}x$
   (e) $2y = 4 - 2x$ (f) $y + 2x = 3$

> **Remember:**
> Perpendicular lines are at right angles to each other.

2 Find an equation of the line perpendicular to $y = 3x - 2$ which goes through

First find the gradient for $y = mx + c$.

   (a) (0, 0)   **(b)** (3, 3)

3 Find an equation of the line perpendicular to $2y = x + 1$ which goes through

   (a) (0, 0)   **(b)** (5, 2)

4 Find an equation of the line perpendicular to $3y + 2x = 7$ which goes through

   (a) (0, 0)   **(b)** (2, −4)

5 Work out the coordinates of the midpoint of the line segments between

   (a) $A(0, 0)$ and $B(5, 6)$ **(b)** $C(1, 1)$ and $D(8, 7)$

   (c) $E(2, 1)$ and $F(5, 4)$ **(d)** $G(3, 2)$ and $H(5, 0)$

6 Work out the coordinates of the midpoint of the line segments between

   (a) $I(-1, 4)$ and $J(3, -5))$ **(b)** $K(2, -3)$ and $L(5, 7)$

   (c) $M(-4, 1)$ and $N(0, 4)$ **(d)** $P(-2, -6)$ and $Q(9, 8)$

Exercise 9.5

Links 9H–J

1 The points $(a, 2)$, $(3, b)$, $(-5, c)$, $(d, 2d)$ lie on the line $4x + y - 3 = 0$. Find a, b, c, d.

2 The points $(-1, a)$, $(b, -3)$, $(2c, 3c)$ lie on the line $2x - 3y + 6 = 0$. Find a, b, c.

3 Use the graph to solve these simultaneous equations

$$x - y = 1$$
$$\tfrac{1}{2}x - y = -1$$

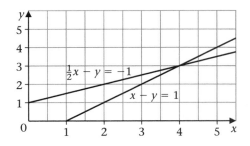

Remember: The intersection of two graphs at (x, y) gives the solutions of x and y.

4 Draw appropriate straight lines to solve these pairs of simultaneous equations.

Remember: Draw the lines by plotting three points.

   (a) $2x + y = 3$
          $x + 3y = 4$

   (b) $5x + 2y = 12$
          $3x - y = 5$

   (c) $4x + y = 14$
          $x - 2y = -1$

   (d) $2x - 3y = 7$
          $x + y = 1$

5 Use an algebraic method to solve the following pairs of simultaneous equations.

Remember: Check your solutions by substituting them into the equations.

(a) $3x + 4y = 10$
$2x + 4y = 8$

(b) $5x + 2y = 22$
$3x + 2y = 10$

(c) $2x + 3y = 4$
$2x - y = -4$

(d) $x + 5y = 8$
$4x + 5y = 2$

(e) $7x + 3y = 9$
$2x - 3y = 18$

(f) $6x - 3y = 21$
$6x + y = 25$

(g) $2x + 5y = 13$
$x + 3y = 2$

(h) $3x - 5y = 9$
$x + y = 2$

(i) $4x - 3y = -17$
$2x + y = 3$

(j) $3x + 2y = 15$
$2x - 3y = 9$

Exercise 9.6   Link 9K

Set up equations and solve.

Write the problem as two simultaneous equations.

1 Tickets for the cinema cost £4 and £5. There were 223 customers who paid a total of £963. How many bought £4 tickets?

2 Boxes of chocolates weighing 200 g and 500 g were packed into a crate. The total weight of the boxes of chocolates was 16 kg and there were 50 boxes in total. How many of each size were there?

3 A string of decorative lights consists of 15 watt and 20 watt bulbs. The total power is 3000 watts (3 kilowatts) and there are 161 bulbs in total. Work out how many of each there are.

4 A father is x years old and his daughter is y years old. Three times the difference in their ages is 57. In 2 years' time the father will be twice as old as his daughter. How old are they now?

5 In a game, 7 reds and 4 blues scores 29 points. 5 reds and 7 blues also scores 29 points. How many points are scored for a red?

6 At a flower shop 23 roses and 21 carnations cost £7.00. 15 roses and 30 carnations cost £7.50. Work out the cost of a single carnation.

7 The line $px - qy = 17$ goes through $(3, -4)$ and $(7, 2)$. Find p and q.

Exercise 9.7   **Link 9L**

1 Draw diagrams to show the regions which satisfy these inequalities.

    (a) $x < 5$ **(b)** $y \geq -3$ **(c)** $-1 < x \leq 2$

    (d) $1\frac{1}{2} \leq y < 2\frac{1}{2}$ **(e)** $x + y < 3$ **(f)** $x - y > 2$

    (g) $3x + y > 0$ **(h)** $y \leq 4x$ **(i)** $x > 2 - 3y$

    (j) $2x + 3y \leq 6$ **(k)** $4y - x > 0$ **(l)** $-2 < 2x + 1 < 5$

2 On a graph shade the region for which

    $x + 2y \leq 6$, $0 \leq x \leq 4$ and $y \geq 0$.

    [E]

> **Remember:** Draw dotted or solid lines depending on the inequality.

3 Draw a diagram to show the region which satisfies

    $y < x + 1$; $y + 2 > 2x$; $3y + 2x > 6$.

4 Sketch the region defined by the inequalities

    $2x = 3y < 12$; $y < 3x$; $x > 2y$.

5 Find the inequalities which define the shaded area.

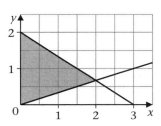

10 Presenting and analysing data 1

Exercise 10.1 Link 10A

1 The manager at 'Fixit Exhausts' records the time, to the nearest minute, to repair the exhausts on 20 cars. Here are his results.

> **Remember:** A stem and leaf diagram should
> • have ordered data values
> • include a key.

32	29	34	28	22	41	57	43	28	33
35	25	52	47	39	27	36	48	53	44

Draw a stem and leaf diagram to show this information.   [E]

2 The stem and leaf diagram shows the marks of some students in a test.

> **Remember:** The mode of a set of data is the value which occurs most often.

```
0 | 5   7
1 | 0   1   1   3   6   7
2 | 1   4   4   5   7   7   8
3 | 0   0   2   2   2            Key: 2|1 = 21
```

Work out the range, the mode and the median. It is likely that the test was marked out of 32. Why?

3 The height of some seedlings is measured to the nearest mm. Half of the seedlings had been fed a special plant supplement. The other half had no special treatment. Here are the stem and leaf diagrams for the results.

```
        Untreated              Treated
    7  6  6  4  3 | 2 | 2  7
 9  8  7  6  5  1 | 3 | 0  3  4
    9  8  5  4  0 | 4 | 1  5  6  9  9
          2  2  2 | 5 | 2  2  3  6  7  7  9
                  | 6 | 0  1
Key: 3|2 = 23                  Key: 3|0 = 30
```

Work out the range, mode and median for the two sets.
Comment on whether the treatment works.

Exercise 10.2 Link 10B

1 The art coursework marks for twelve students marked out of 80 were

> 51, 54, 57, 38, 63, 49, 54, 72, 76, 28, 46, 54

(a) Work out the
(i) mean **(ii)** median **(iii)** mode
for these marks.

(b) Comment on your results.

(c) The moderator raises all the marks by 4. Work out the mean, median and mode of the new marks.

> **Remember:** The mean of a set of data is the sum of the values divided by the number of values. The median is the middle value when the data is arranged in order of size.

2 The diagram represents a wheel which is spun about the centre.
The number the arrow points to is the number recorded.
The spinner is spun several times.

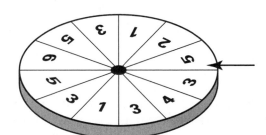

(a) What is the theoretical mean of the recorded numbers?

(b) Explain your result.

(c) 'The mean is not a sensible measure of average.' Explain why.

(d) What are the mode and median of the numbers? Explain your results.

3 Sebastian wishes to conduct a survey on the amount of pocket money teenagers receive. Explain, with reasons, which averages he should use and why.

Exercise 10.3 Links 10C–F

1 Here are the numbers of seats won by the Conservative party in General Elections from 1951 to 2005.

> 321 345 365 304 253 330 297 277
> 339 397 376 336 165 166 198

(a) Work out the median, lower quartile and upper quartile.

(b) Work out the range and interquartile range.

> Order the values.
> The lower quartile is the
> $\dfrac{(15 + 1)}{4} = $ 4th data value.
> The upper quartile is the
> $\dfrac{3(15 + 1)}{4} = $ 12th data value.

2 The weights of packets of sugar are being checked to ensure that the packing machinery is working properly. The difference between the actual and stated values is recorded. Overweight is positive and underweight is negative.

> +1.1 g, +2.1 g, −0.5 g, +1.3 g, +0.9 g, −0.3 g,
> +1.3 g, +1.9 g, −0.4 g, +1.2 g, +2.0 g

Work out the interquartile range and the median.

3 The table shows the number of text messages received by each of 31 students one evening.

> The smallest value is 0 and the largest value is 5.

Number of text messages	0	1	2	3	4	5
Frequency	2	4	7	8	7	3

Work out the median, range and interquartile range.

4 The table is a record of the numbers of photographs taken on rolls of film which are described as 36 exposure.

Number of pictures	34	35	36	37	38
Frequency	1	3	26	31	4

Work out the mean, mode and median.

> Mean $= \dfrac{\Sigma fx}{\Sigma f}$
>
> $= \dfrac{1 \times 34 + 3 \times 35 + \cdots}{65}$

5 The bus fares in London are set according to the number of zones travelled through. The table shows the fares payable and the numbers of passengers purchasing these tickets.

> **Remember:** The mode is the value with the highest frequency.

Fare	80p	£1.20	£2.50	£3.00
Frequency	103	72	23	12

Work out the mean, mode and median.

6 The lifetimes in hours of 100 batteries are shown in the table.

Time t (hours)	Frequency
$0 \leqslant t < 3$	2
$3 \leqslant t < 6$	3
$6 \leqslant t < 9$	16
$9 \leqslant t < 12$	34
$12 \leqslant t < 15$	27
$15 \leqslant t < 18$	13
$18 \leqslant t < 21$	5

> Use the midpoints of the intervals.
>
> So mean $= \dfrac{\Sigma fx}{\Sigma f}$
>
> $= \dfrac{2 \times 1.5 + 3 \times 4.5 + \cdots}{100}$

(a) Calculate an estimate of the mean.

(b) Write down the class interval in which the median lies.

(c) State the modal class.

> **Remember:** The modal class is the class interval with the highest frequency.

7 Andrew did a survey at the seaside for his science coursework. He measured 55 pieces of seaweed. The results of his survey are shown in the table.

Length of seaweed (cm)		Frequency	
$0 < L \leqslant 20$		2	
$20 < L \leqslant 40$		22	
$40 < L \leqslant 60$		13	
$60 < L \leqslant 80$		10	
$80 < L \leqslant 100$		5	
$100 < L \leqslant 120$		2	
$120 < L \leqslant 140$		1	

(a) Work out an estimate for the mean length of the pieces of seaweed. Give your answer to 1 decimal place.

(b) Write down the interval which contains the median length of a piece of seaweed. [E]

Exercise 10.4

Link 10G

1 The Year 11 pupils at Harkspar High School sat an exam in geography. The exam consisted of two papers, A and B. The table shows the medians and interquartile ranges of the marks on the two papers.

Paper	Median (%)	Interquartile range
A	61	16
B	82	23

Comment on this data.

2 The table shows the medians and interquartile ranges of the weights of two rugby teams.

Team	Median (kg)	Interquartile range
A	89	17
B	95	14

Explain whether or not it would be fair to say that members of team B are in general heavier than members of team A.

3 In an examination consisting of two papers, the following statistics were collected.

Paper	Median % mark	Interquartile range
1	54	28
2	71	15

Make statistical comparisons between Paper 1 and Paper 2.

Exercise 10.5 Link 10H

1 Alicia has information about the quarterly gas bill at her home over a four-year period from 1998 to 2001. The information is given in the table below.

YEAR	February	May	August	November
1998	252	203	112	160
1999	264	210	118	163
2000	273	215	124	172
2001	294	224	131	189

> The first four-point moving average is
> $$\frac{252 + 203 + 112 + 160}{4} = 181.75.$$
> The second four-point moving average is
> $$\frac{203 + 112 + 160 + 264}{4} = 184.75.$$
> ... and so on.

(a) Represent this data as a time series.

(b) Calculate the four-point moving averages and plot these on the same axes as the original data.

(c) Complete the trend line for the moving averages.

(d) Comment as fully as possible on the changes in the gas bill over the four-year period.

2 The average house prices (£1000s) in Shimpwell from 1985 to 2001 are given below.

Year	85	86	87	88	89	90	91	92	93	94	95	96	97	98	99	00	101
Price	47	47	56	63	48	48	49	48	52	54	57	62	68	75	90	98	105

(a) Plot the data as a time series.

(b) Work out the 5-point moving averages, plotting these as the trend line on the same axes as the time series.

(c) Comment on the data and your graphs.

Exercise 10.6

Links 10I, J

1   The marks obtained by 15 pupils in two maths tests are shown in the table.

Test 1	94	90	52	72	60	75	63	70	40	81	59	48	66	78	75
Test 2	80	86	40	74	54	70	63	65	44	78	62	46	62	74	80

(a)   Draw a scatter diagram.

(b)   Draw on the line of best fit.

> **Remember:** The line of best fit is drawn to be as close as possible to all the plotted points on the scatter diagram.

(c)   Comment on the type of correlation.

(d)   Use your line of best fit to estimate a mark for Test 1 for a pupil who scores 58 on Test 2.

(e)   Use your line of best fit to estimate a mark for Test 2 for a pupil who scores 32 on Test 1.

> **Remember:** When the points on a scatter diagram are very nearly along a straight line there is a high correlation between the variables.

2   The table shows the minimum temperature in January and the height above sea level for ten towns.

Height above sea level (m)	200	500	700	850	1500	2000	3800	4000	4500	4800
Jan minimum temp (°C)	4	7	5	−1	−4	−3	−10	−8	−12	−18

(a)   Draw a scatter diagram for this data.

(b)   Comment on the type of correlation.

(c)   Estimate the minimum temperature in January for a major town whose height above sea level is 1200 m.

3   Ten pupils took two examination papers in science. Their marks, out of 100, were

Paper 1	88	48	80	96	60	50	20	74	78	68
Paper 2	86	56	76	84	64	60	50	70	80	74

(a)   Draw a scatter diagram for these marks.

(b)   Draw on the line of best fit.

(c)   Comment on the type of correlation.

(d)   Use your line of best fit to estimate the mark given for Paper 2 to a pupil who scores 64 on Paper 1.

(e)   Work out the equation for the line of best fit.

11 Estimation and approximation

Exercise 11.1

Link 11A

1 Round these numbers to the nearest 10.

    (a) 379   **(b)** 438   **(c)** 94   **(d)** 5999

2 Round these to the nearest whole number.

    (a) 4.6   **(b)** 0.499   **(c)** 11.85   **(d)** 5.515

3 Rewrite these with each number rounded to the nearest whole number.

    (a) 3.8×7.4 **(b)** 0.51×6.98 **(c)** 8.6×3.9

4 54 498 people watched a football match.

    (a) Write this number to the nearest

        (i) 10   **(ii)** 100   **(iii)** 1000   **(iv)** 10 000

    (b) Which figure is it most sensible to use in a newspaper headline and why?

5 Write these numbers correct to the approximation given in brackets.

    (a) 0.054926   (4 d.p.)
    (b) 458.349   (1 d.p.)
    (c) 50.0509   (3 d.p.)
    (d) 9.999   (2 d.p.)

6 Carry out the following calculations. Give your answers correct to the number of decimal places given in brackets.

    (a) 18.47×1.563 (2 d.p.)
    (b) 3.142×9.666 (3 d.p.)
    (c) 19.698×25.923 (1 d.p.)

Exercise 11.2

Links 11B, C

1 Write these numbers correct to the approximation given in brackets.

    (a) 40.49   (3 s.f.)
    (b) 99.98   (2 s.f.)
    (c) 0.04543   (3 s.f.)
    (d) 9.0909   (2 s.f.)
    (e) 0.00467   (1 s.f.)
    (f) 104.9   (2 s.f.)
    (g) 1.0449   (3 s.f.)

> **Remember:** Count to the required number of places.
> Look at the next place:
> 5 or more → Round up
> 4 or less → Leave alone

> **Remember:** Count the required number of significant figures.
> Ignore 0s at the front.
> Look at the next digit.
> 5 or more → Round up
> 4 or less → Leave alone
> Check that the number keeps its size.

2 68 726 people attended a pop concert at Woburn.
Write this number correct to
   (a) 1 s.f.   **(b)** 3 s.f.   **(c)** 2 s.f.

 3 Work out the following calculations. Give your answers correct to
the number of significant figures given in brackets.
   (a) 439×49 (2 s.f.)
   (b) 4.8×0.313 (4 s.f.)
   (c) $976 \times 453 \times 182$ (1 s.f.)
   (d) 0.00149×0.0837 (3 s.f.)
   (e) $77 \times 88 \times 99$ (3 s.f.)

4 1 centimetre is approximately equal to 0.394 inches. Write
   (a)   1 foot (12 inches) in centimetres correct to 2 s.f.
   (b)   1 yard (3 feet) in centimetres correct to 3 s.f.
   (c)   1 mile (1760 yards) in centimetres correct to 1 s.f.

5 In each of parts **(a)** to **(d)**
   (i)   write down a calculation that could be used to estimate the
            answer
   (ii) work out the estimated answer
   (iii) use a calculator to work out the exact answer.

> Write each calculation to
> 1 s.f. first.

   (a) 3.42×7.69 **(b)** $(1268 \times 176) + 402$
   (c) $\dfrac{259.4 \times 46.9}{83.2}$ **(d)** $\dfrac{0.0456 \times 0.00296}{0.453}$

6 Estimate the answer to each of the following. Give your answers
to the nearest whole number.
   (a) $41 \div 8$ **(b)** $111 \div 12$
   (c) $198 \div 43$ **(d)** $287 \div 21$
   (e) $497 \div 36$ **(f)** 142×27
   (g) $\dfrac{46 \times 33}{78}$ **(h)** $\dfrac{54 \times 71}{36}$

7 Michael is working out the radius of a circle.
He has to do the calculation

$$r = \sqrt{\dfrac{76.99}{3.142}}$$

Estimate, to the nearest whole number, r, the radius of the circle.

Exercise 11.3 **Links 11D, E**

1 Copy and complete this table showing the attendance to the nearest hundred at these rugby stadiums.

Rugby stadium	Attendance to the nearest 100	Lowest possible attendance	Highest possible attendance
Newcastle	1200		
Bedford	4600		
Saracens	11 700		
Leicester	13 200		
Bath	9800		

> **Remember:** When **adding** two measurements $a + b$
> Add upper bounds to get the maximum value
> Add lower bounds to get the minimum value
> When **subtracting** two measurements $a - b$
> Subtract lower bound b from upper bound a to get the maximum value
> Subtract upper bound b from lower bound a to get the minimum value
> When **multiplying** two measurements $a \times b$
> Multiply upper bounds to get the maximum value
> Multiply lower bounds to get the minimum value
> When **dividing** two measurements $a \div b$
> Divide upper bound a by lower bound b to get the maximum value
> Divide lower bound a by upper bound b to get the minimum value

2 The distance between Hitchin and Shefford is given as 8 miles on a road sign.
Write down the range in which the true distance could lie.

3 These lengths are given to the nearest cm.
Write down the smallest and largest possible lengths they could be.

    (a) 48 cm   **(b)** 26.43 m   **(c)** 9.05 m   **(d)** 1.3842 km

4 The length of each side of a regular octagon is 5.6 cm correct to 2 significant figures. Calculate the greatest length the perimeter could be.

5 The speed of light is 186 000 miles per second to the nearest thousand miles per second.
The distance of the Earth from the Sun is 93.5 million miles correct to the nearest hundred thousand miles.
A ray of light leaves the Sun and travels to the Earth. It takes time T.
Calculate the range within which the time T taken by the ray lies.

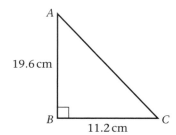

6 *ABC* is a right-angled triangle.

    (a) Write down
        (i) the upper bound of *BC*   **(ii)** the lower bound of *BC*.

    (b) Calculate
        (i) the upper bound of the area of the triangle
        (ii) the lower bound of the area of the triangle.

12 Sequences and formulae

Exercise 12.1
Links 12A, B

In questions **1–6** find the first five terms in the sequence.

1 $u_n = 3n$ **2** $u_n = n + 4$ **3** $u_n = 2n^2$

4 $u_n = 2n + 1$ **5** $u_n = n^2 - 1$ **6** $u_n = (2n + 1)^2$

In questions **7–15** find the value of n for which u_n has the stated value.

7 $u_n = 3n - 1$ $u_n = 59$ **8** $u_n = 6n - 3$ $u_n = 45$

9 $u_n = n^2 + 3$ $u_n = 124$ **10** $u_n = 2n^2$ $u_n = 162$

11 $u_n = n^3$ $u_n = 1728$ **12** $u_n = n(n + 1)$ $u_n = 42$

13 $u_n = \dfrac{2n + 1}{3n - 1}$ $u_n = \frac{7}{8}$ **14** $u_n = \dfrac{4 - n}{8 - 2n}$ $u_n = \frac{3}{2}$

15 $u_n = 2^{(n - 4)}$ $u_n = 512$

Remember: Another way of writing a sequence is
$u_n = 2n - 1$, where u_n is the nth term and n is the term number.

Term number	Term
1	1 ⎫ +2
2	3 ⎬ +2
3	5 ⎭
...	...
n	$2n - 1$

The 2 goes here.

Exercise 12.2
Link 12C

Find the formulae for u_n to describe each of the following.

1   1, 4, 7, 10, 13 ...   **2**   3, 5, 7, 9, 11 ...

3   10, 15, 20, 25, 30 ...   **4**   −3, −1, 1, 3 ...

5   17, 14, 11, 8, 5 ...   **6**   3, $3\frac{1}{2}$, 4, $4\frac{1}{2}$, 5 ...

7   0, 3, 8, 15, 24 ...   **8**   3, 9, 19, 33, 51 ...

9   4, 12, 36, 108, 324 ...   **10**   2, 10, 50, 250, 1250 ...

Exercise 12.3
Links 12D, E

1 $A = LB$ Calculate the value of A when $L = 2.7$ and $B = 5.2$.

Substitute the values into the formula.

2 $P = 2(L + B)$ Calculate the value of P (correct to 2 s.f.) when $L = 1.07$ and $B = 4.23$.

3 $A = 2\pi r(r + h)$ Calculate the value of A (correct to 3 s.f.) when $r = 5.72$ and $h = 6.11$.

4 $x = L(1 + at)$ Calculate the value of x (correct to 2 s.f.) when $L = 120$, $a = 0.000\,02$ and $t = 50$.

5 $v = u + at$ Calculate the value of v when $u = 35$, $a = 6$ and $t = 5$.

6 $I = \dfrac{PRT}{100}$ Calculate the value of I when $P = 2500$, $R = 6.25$ and $T = 2$.

7 $h = ut + \frac{1}{2}at^2$ Calculate the value of h (correct to 1 d.p.) when $u = 63.1$, $a = -9.81$ and $t = 3$.

8 $f = \dfrac{u + v}{uv}$ Calculate the value of f (correct to 2 s.f.) when $u = 6.4$ and $v = 3.6$.

9 $T = 2\pi\sqrt{\dfrac{l}{g}}$ Calculate (correct to 3 s.f.) the value of T when $l = 16.2$ and $g = 9.82$.

10 $T = \dfrac{2Mmg}{M + m}$ Calculate (correct to 2 s.f.) the value of T when $M = 5.6$, $m = 12$ and $g = 9.8$.

11   Matthew uses this formula to calculate the value of D.

$$D = \frac{a - 3c}{a - c^2}$$

Calculate the value of D when $a = 19.9$ and $c = 4.05$.   [E]

12 $y = ab + c$
Calculate the value of y when $a = \frac{3}{4}$, $b = \frac{7}{8}$ and $c = -\frac{1}{2}$.   [E]

13   The volume, V, of a barrel is given by the formula

$$V = \pi H(2R^2 + r^2) \div 3000$$

$\pi = 3.14$, $H = 60$, $R = 25$ and $r = 20$.
Calculate the value of V correct to 3 significant figures.   [E]

Exercise 12.4   Links 12F, G

1   Make the letter in brackets the subject of the formula.

(a)	$x = a(b + c)$	$[a]$		(b)	$x = a(b + c)$	$[b]$
(c)	$M = DV$	$[V]$		(d)	$E = mc^2$	$[c]$
(e)	$s = \frac{1}{2}gt^2$	$[g]$		(f)	$s = \frac{1}{2}gt^2$	$[t]$
(g)	$x = L(1 + at)$	$[L]$		(h)	$x = L(1 + at)$	$[a]$
(i)	$s = ut + \frac{1}{2}at^2$	$[a]$		(j)	$p = qr + q$	$[q]$

Remember: Making a letter the subject of a formula is a bit like solving an equation. You use the same skills.

2   For the formula $W = \dfrac{\lambda x^2}{2L}$ make

(a) L the subject and   (b) x the subject.

3 $A = LB$ Calculate B when $A = 32$ and $L = 4$.

4 $P = 2(L + B)$ Calculate L (correct to 2 s.f.) when $P = 26.4$ and $B = 3.1$.

5 $A = 2\pi r(r + h)$ Calculate h (correct to 3 s.f.) when $A = 122$ and $r = 4$.

6 $x = L(1 + at)$ Calculate L (correct to 2 s.f.) when $x = 15.63$, $a = 0.000\,014$ and $t = 60$.

7 $r = u + at$
    (a) Calculate a when $v = 40$, $u = 24$ and $t = 2$.
    (b) Calculate t when $v = 14$, $u = 46$ and $a = -5$.

8 $I = \dfrac{PRT}{100}$ Calculate P when $I = 62.5$, $R = 5$ and $T = 3$.

9 $h = ut + \frac{1}{2}at^2$
    (a) Calculate u (correct to 1 d.p.) when $h = 12$, $t = 1$ and $a = 4$.
    (b) Calculate a (correct to 1 d.p.) when $h = 15.6$, $t = 3$ and $u = 4.2$.

10 $T = 2\pi\sqrt{\dfrac{l}{g}}$ Calculate l when $T = 0.45$ and $g = 9.8$.

11 $T = \dfrac{2Mmg}{M + m}$
    (a) Calculate g when $T = 23.4$, $M = 30$ and $m = 20$.
    (b) Calculate M when $T = 7.6$, $g = 9.8$ and $m = 36$.

Exercise 12.5   **Link 12H**

1 $E = fd$ and $f = ma$
Find a formula for E in terms of m, a and d.

2 $s = ut + \frac{1}{2}at^2$ and $v = u + at$
Find a formula for s in terms of t, u and v.

3 $x = 3a^2$ and $y = 6a$
Find a formula for x in terms of y.

4 $y = rv^2$ and $v = \dfrac{r}{w}$

Find a formula for y in terms of r and w.

5 $x = st^2$ and $v = \dfrac{t}{s}$

Find a formula for x in terms of s and v.

⑬ Measure and mensuration

Exercise 13.1

Links 13A, B

1 A table mat measures 25 cm by 20 cm. Its dimensions are quoted to the nearest centimetre.

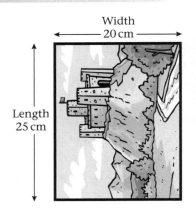

Width
←— 20 cm —→

Length
25 cm

Remember:
Area of rectangle
= length × width
= $l \times w = lw$
Perimeter of rectangle
= $2(l + w)$

    (a) Write down the shortest and longest possible width of the mat.

    (b) Write down the shortest and longest possible length of the mat.

    (c) Calculate the smallest and largest possible area of the mat.

2

$(2x + 1)$ cm

$(x + 3)$ cm

    (a) Find in its simplest form an expression in x for the perimeter of this rectangle.

    (b) Given that the perimeter is 104 cm, calculate the value of x.

    (c) Calculate the area of the rectangle.

3 The length of a rectangle is twice the width. The area of the rectangle is 162 cm². Calculate the perimeter of the rectangle.

4 Calculate the areas of these triangles.

    (a)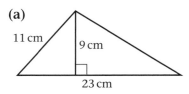

11 cm   9 cm
23 cm

    (b)

8 cm
←6.5 cm→

Remember:

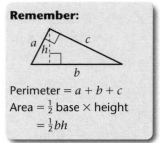

a h c
b

Perimeter = $a + b + c$
Area = $\frac{1}{2}$ base × height
= $\frac{1}{2}bh$

5 The perimeter of this triangle is 32 cm.

    (a) Calculate the value of x.

    (b) Write down the lengths of the three sides.

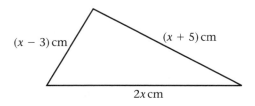

$(x - 3)$ cm $(x + 5)$ cm
$2x$ cm

6 Calculate the areas of these parallelograms.

(a)

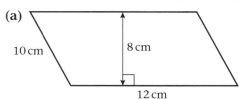

(b)

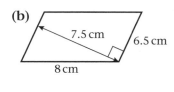

Remember:

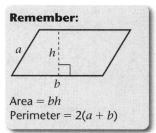

Area = bh
Perimeter = $2(a + b)$

7 The diagram shows a rectangle and a parallelogram.

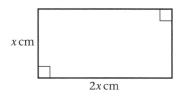

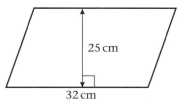

These two shapes have equal areas.

(a) Calculate the value of x.

(b) Write down the lengths of the two sides of the rectangle.

8 Calculate the areas of these trapeziums.

(a)

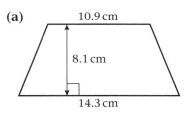

(b)

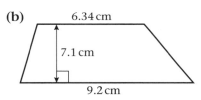

Remember:

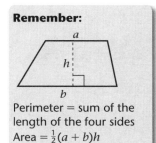

Perimeter = sum of the length of the four sides
Area = $\frac{1}{2}(a + b)h$

9 Find the shaded areas.

(a)

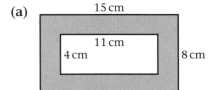

(b)

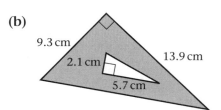

Exercise 13.2

Link 13C

1 Calculate the circumferences and areas of these circles.

(a)

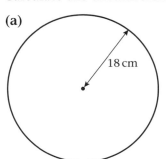

(b)

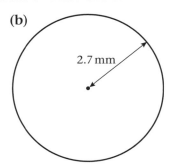

Remember:

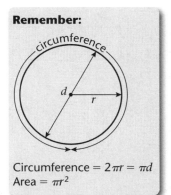

Circumference = $2\pi r = \pi d$
Area = πr^2

2 The area of a circle is 150 cm².
Calculate the circumference of the circle.

3 The circumference of a circle is 68 cm.
Calculate the area of the circle.

4 The diagram represents a running track.
It consists of a rectangle and two semicircles.

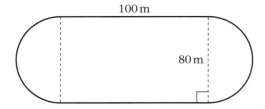

100 m

80 m

Calculate

(a) the perimeter of the running track

(b) the area enclosed by the running track.

Exercise 13.3 **Link 13D**

1 Calculate the volume of a cube of side length

(a) 6 cm

(b) 4.3 cm

(c) 29.8 mm

> **Remember:** For a cube of edge length a
> Volume $= a^3$

2 A cube has a volume of 729 cm³.
Calculate

(a) the length of the side of the cube

(b) the surface area of the cube.

3 Calculate the volumes and surface areas of these cuboids.

(a)

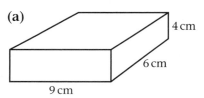

4 cm

6 cm

9 cm

(b)

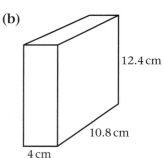

12.4 cm

10.8 cm

4 cm

> **Remember:**
>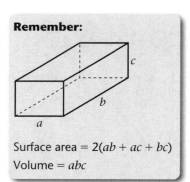
> c
> b
> a
> Surface area $= 2(ab + ac + bc)$
> Volume $= abc$

4 Calculate the volumes and surface areas of these prisms.

(a)

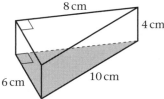

(b)

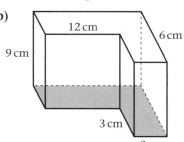

Remember:
For a prism
Volume = area of base
$\times$ vertical height

5 The prism *ABCDEF* has volume = 510 cm³,
BC = 10 cm, *DC* = 12 cm.
Calculate the length *AB*.

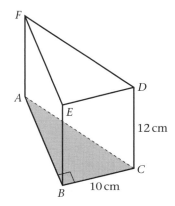

Exercise 13.4

Link 13E

1 Here are two expressions related to a cone of radius *r*, height *h* and slant height *l*.

$$\pi r l \quad \text{and} \quad \tfrac{1}{3}\pi r^2 h$$

State, with reasons, which expression gives the volume of the cone and which gives the curved surface area.

Remember:
Length has dimension 1,
area has dimension 2,
volume has dimension 3.

2 In the following expressions λ is dimensionless and *a*, *b*, *c* have dimensions of length.
Explain whether each expression represents a length, an area or a volume.

(a) $a\lambda^2 b$ **(b)** $abc\lambda$ **(c)** λa **(d)** $\lambda a^2 c$ **(e)** $\lambda(a + b + c)$

3 Here are three expressions.

Expression	Length	Area	Volume	None of these
πxyz				
$\pi xy + \pi y^2$				
$\pi x + \pi yx$				

x, *y* and *z* are lengths. π has no dimensions.
Copy the table and put a tick ($\checkmark$) in the correct column to show whether the expression can be used for length, area, volume or none of these.

14 Simplifying algebraic expressions

Exercise 14.1

Links 14A, B

Simplify

1 $(x^3)^2$

2 $(xy)^3$

3 $(x^2yz^3)^2$

4 $a^3 \times 3a$

5 $5a^2 \times 2a^3$

6 $a^2b^3 \times ab$

7 $abc \times ab$

8 $a^4 \times b^3c \times ab$

9 $(2x^2y)^3$

10 $2pq(p^2q + pq^2)$

11 $x^5 \div x^2$

12 $x^4 \div x$

13 $4x^6 \div 2x^3$

14 $a^2bc \div ab$

15 $15x^2z \div 3xz$

16 $9p^3q^4 \div 3p^2q^2$

17 $c^3 \div c$

> **Remember:**
> $x^m \times x^n = x^{m+n}$
> $(x^m)^n = x^{mn}$
> $x^m \div x^n = x^{m-n}$

Exercise 14.2

Links 14C–E

Find the value of

1 $27^{\frac{1}{3}}$

2 $(\frac{4}{9})^{\frac{1}{2}}$

3 $(343)^{\frac{1}{3}}$

4 $(36)^{-\frac{1}{2}}$

5 $(121)^{-\frac{1}{2}}$

6 $(32)^{\frac{1}{5}}$

7 $(216)^{\frac{2}{3}}$

8 $9^{\frac{5}{2}}$

9 $(512)^{-\frac{2}{3}}$

10 $(\frac{9}{4})^{-\frac{3}{2}}$

11 $(\frac{1}{16})^{-\frac{1}{2}}$

12 $(\frac{36}{49})^{\frac{3}{2}}$

13 16^0

14 $(\sqrt{8})^{\frac{1}{3}}$

15 $(81)^{\frac{3}{4}} \div (9)^{\frac{1}{2}}$

> **Remember:**
> $x^{\frac{1}{2}} = \sqrt{x}$
> $x^{\frac{1}{n}} = \sqrt[n]{x}$
> $x^{\frac{m}{n}} = \sqrt[n]{x^m}$
> $x^{\frac{m}{n}} = (\sqrt[n]{x})^m$

Find the values of k.

16 $x^{2k} = (x^3)^{\frac{1}{2}}$

17 $x^k = 1 \div x^{-1}$

18 $y^{k+1} = (y^3)^3$

19 $(\sqrt{y})^k = y$

20 $729^k = 9$

21 $16^k = 64$

22 $4^{\frac{k}{2}} = 32$

> **16:** $2k = 3 \times \frac{1}{2}$

Simplify

23 $(4a^2)^{\frac{1}{2}}$

24 $(27a^6b^3)^{\frac{1}{3}}$

25 $(\sqrt{2}x)^4$

26 $(\sqrt{3}y^2)^2$

27 $(a^{-2}b)^{-1}$

28 $(x^{-4}y^2)^{-\frac{1}{2}}$

29 $9x\sqrt{x}(3x)^{-\frac{3}{2}}$

30 $(16x^2y^{-2})^{-\frac{1}{2}}$

31 $(8x^6)^{\frac{1}{3}} + (4x^4)^{\frac{1}{2}}$

Exercise 14.3

Links 14F, G

Simplify

1 $\dfrac{4xy}{2y}$

2 $\dfrac{9x^2}{3x}$

3 $\dfrac{18xy^2}{3xy}$

> Cancel the common factors.

4 $\dfrac{24a^2bc}{9abc^2}$

5 $\dfrac{2(x+2)}{y(x+2)}$

6 $\dfrac{2b^2}{a} \times \dfrac{a^3}{4b}$

7 $\dfrac{(x+1)(x-1)}{2(x-1)}$

8 $\dfrac{3x^2}{6(x+1)}$

9 $\dfrac{(z-2)^3}{(z-2)(z+2)}$

Write as a single fraction

10 $\dfrac{1}{4} + \dfrac{2}{7}$

11 $\dfrac{x}{4} + \dfrac{2x}{7}$

12 $\dfrac{2x}{9} + \dfrac{x}{3}$

> **Remember:**
> $\dfrac{1}{n} + \dfrac{1}{m} = \dfrac{m+n}{mn}$

13 $\dfrac{1}{4} + \dfrac{1}{2x}$

14 $\dfrac{a}{3} + \dfrac{b}{5}$

15 $\dfrac{2c}{7} - \dfrac{3b}{5}$

16 $\dfrac{p+2}{3} + \dfrac{p-1}{5}$

17 $\dfrac{2q+3}{2} + \dfrac{3q-4}{3}$

18 $\dfrac{2}{y} - \dfrac{3}{4y}$

19 $\dfrac{1}{x+1} + \dfrac{1}{x+2}$

20 $\dfrac{2}{p-3} - \dfrac{5}{p+1}$

21 $\dfrac{a}{a+3} - \dfrac{3}{a+1}$

Write down the LCM for these denominators.

22 15, 40

23 74, 111

24 x^3, x^4

> **Remember:** LCM means lowest common multiple.

25 $xz, 2x^2$

26 $3(x+1), 5(x-1)$

Work out as a single fraction in its lowest terms.

27 $\dfrac{b+1}{bc} + \dfrac{1-a}{ac}$

28 $\dfrac{1}{a(b+1)} - \dfrac{1}{b(a+1)}$

29 $\dfrac{2x}{y-1} - \dfrac{x}{y}$

30 $\dfrac{3x}{2x+1} + \dfrac{1}{3}$

31 $\dfrac{x-1}{x+1} + \dfrac{x+1}{x-1}$

32 $\dfrac{1}{(x-1)^2} - \dfrac{1}{(x-1)}$

Exercise 14.4

Links 14H–J

Factorise completely

1 $6x^2 + 2x$

2 $3a^2b + 3ab$

3 $xy^2 - xy$

4 $6c^2 - 4c^4$

5 $4a^2 + 2ac$

6 $8a^3 + 2a^2b$

7 $14a^3b + 21ab^2$

8 $3x^2y - 6xy^2 + 9xy$

> Look for common factors to take outside the bracket.

9 $15x^4 - 3x^2$

10 $5x^2y - 10xy^2$

11 $y(x + y) + a(x + y)$

12 $b(2x - y) - c(2x - y)$

13 $(2x + 1)(a + b) + (1 - x)(a + b)$

14 $3c(2a + b) - d(b + 2a)$

15 $2p(3 - q) + 3(q - 3)$

16 $ax + ay + bx + by$

17 $x^2 + xy + y^2 + yx$

18 $2pqr + 2pa - qr - a$

19 $x^2 + 3x - 15x - 45$

20 $x^2 + 4x + 8x + 32$

21 $x^2 - 24 - 6x + 4x$

22 $x^2 - 11x + 18$

23 $x^2 - 9x + 18$

24 $x^2 + 19x + 18$

25 $x^2 - 5x - 24$

26 $x^2 + 10x - 24$

27 $x^2 + 19x + 60$

28 $x^2 + 17x + 60$

29 $x^2 + 2x - 15$

30 $2x^2 - 7x - 15$

31 $3x^2 - x - 4$

32 $6x^2 - 19x - 7$

33 $12x^2 - 7x + 1$

34 $12x^2 - x - 6$

35 $10x^2 - 91x + 9$

36 $5x^2 + 44x - 9$

37 $5x^2 - 12x - 9$

38 $9 - 21x + 10x^2$

39 $x^4 + 3x^2 + 2$

40 $x^{2n} - 5x^n + 6$

> **Remember:** To factorise these quadratics you must end up with two brackets multiplied together, e.g.
> $2x^2 + x - 3$
> $= (2x + 3)(x - 1)$

41 **(a)** Simplify $y^3 \times y^4$.

(b) Expand and simplify $5(2x + 3) - 2(x - 1)$.

(c) Factorise **(i)** $4a + 6$ **(ii)** $6p^2 - 9pq$

(d) Find the value of **(i)** 10^{-2} **(ii)** 7

The number 1104 can be written as $3 \times 2^c \times d$, where c is a whole number and d is a prime number.

(e) Work out the value of c and the value of d.　　　　[E]

Exercise 14.5 **Links 14K, L**

Factorise completely

1 $a^2 - 16$

2 $4y^2 - 81$

3 $36 - 9k^2$

4 $2x^2 - 50$

5 $5y^2 - 80$

6 $8z^2 - 162$

7 $x^2 + 16x + 64$

8 $x^2 - 8x + 16$

9 $2x^2 + 20x + 50$

10 $3x^2 - 36x + 108$

> **Remember:** To factorise the difference of two squares use
> $x^2 - y^2 = (x - y)(x + y)$

> Use
> $x^2 + 2ax + a^2 = (x + a)^2$
> or
> $x^2 - 2ax - a^2 = (x - a)^2$

Simplify

11 $x^2 - (x - 1)^2$

12 $(x + 2)^2 - (x + 1)^2$

13 $101^2 - 99^2$

14 $(2y + z)^2 - (2y - z)^2$

> Factorise, then cancel the the common factors, top and bottom.

Simplify as fully as possible

15 $\dfrac{x - 2}{x^2 - 4}$

16 $\dfrac{3x + 9}{x^2 - 9}$

17 $\dfrac{2x + 8}{x - 3} \times \dfrac{3x - 9}{x + 4}$

18 $\dfrac{x^2 - 49}{2x + 5} \div \dfrac{4x + 28}{4x^2 - 25}$

19 $\dfrac{1}{3x - 2} - \dfrac{x + 1}{3x^2 + 16x - 12}$

20 $\dfrac{x^2 - 5x + 6}{x^2 - 4} \times \dfrac{x^2 + 2x - 3}{x^2 - 9}$

21 $\dfrac{6x^2 + 13x + 5}{4x^2 - 4x - 3}$

22 $\dfrac{x^2 - 7x + 6}{x^2 - 5x - 6}$

23 Write as a single fraction.

$$\dfrac{3}{x + 1} + \dfrac{5x}{2x^2 + 7x + 5}$$

[E]

15 Pythagoras' theorem

Exercise 15.1   **Links 15A–C**

1 Find the length of the hypotenuse in these right-angled triangles.

(a)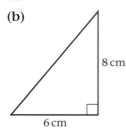
4 cm
3 cm

(b)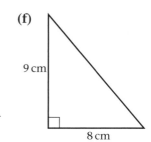
8 cm
6 cm

(c)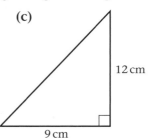
12 cm
9 cm

Remember:

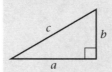
c   *b*
a

The hypotenuse is the side opposite the right angle.
To find the length of the hypotenuse:
$$a^2 + b^2 = c^2$$
Square the sides, add the squares together and then find the square root.

(d)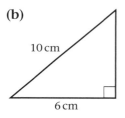
7 cm
6 cm

(e)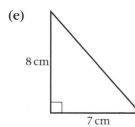
8 cm
7 cm

(f)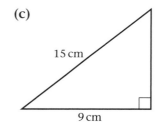
9 cm
8 cm

2 Calculate the length of the unmarked side in these right-angled triangles.

(a)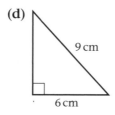
5 cm
3 cm

(b)
10 cm
6 cm

(c)
15 cm
9 cm

Remember:

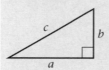
c   *b*
a

To find the length of the shorter sides:
$$a^2 = c^2 - b^2$$
$$b^2 = c^2 - a^2$$
Square, subtract and then find the square root.

(d)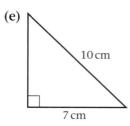
9 cm
6 cm

(e)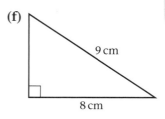
10 cm
7 cm

(f)
9 cm
8 cm

3 A ladder is resting against the wall of a house. The foot of the ladder is 3 m from the base of the wall and the top of the ladder is 4 m from the base of the wall. How long is the ladder?

4 Calculate the span of the roof truss shown in the diagram.

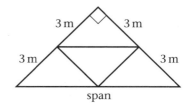

span

5 Calculate the lengths of the sides marked by a letter.

(a)

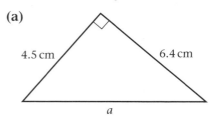

4.5 cm   6.4 cm

a

(b)

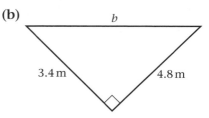

b

3.4 m   4.8 m

(c)

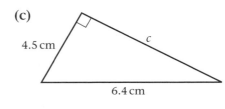

4.5 cm   *c*

6.4 cm

(d)

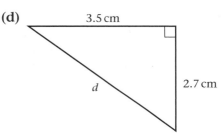

3.5 cm

d   2.7 cm

(e)

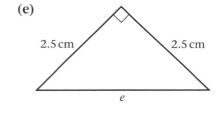

2.5 cm   2.5 cm

e

(f)

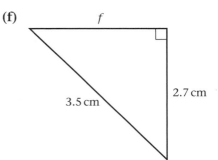

f

3.5 cm   2.7 cm

(g)

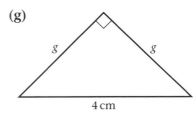

g   *g*

4 cm

6 Susie is flying her kite on a horizontal playing field. The string is taut and the kite is 10 m above the ground. The kite is 30 m from Susie in a horizontal direction. How long is the kite string?

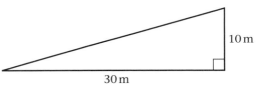

10 m

30 m

7 Susie's friend Sophie is flying her kite on the horizontal playing field. The 30 m length of string on her kite is taut and the kite is 20 m away from Sophie in a horizontal direction. How far from the ground vertically is the kite?

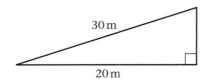
30 m
20 m

Exercise 15.2   **Links 15D–G**

1 Keith used his 6-metre ladder to clean his upstairs windows. He placed the ladder 2 metres away from the foot of the wall. How far up the wall did the ladder reach?

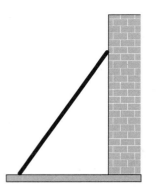

2 Meg used her 8-metre ladder to paint her upstairs windows. She placed the top of the ladder 6 metres above the ground. How far away from the base of the wall was the foot of the ladder?

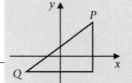
Sketch the axes first.

3 Calculate the distance between each pair of points.

(a) (2, 3) and (5, 8)   (b) (8, 3) and (5, 0)

(c) (−2, 3) and (5, 0)   (d) (−3, −2) and (−5, −5)

Then sketch a right-angled triangle to find the hypotenuse.

4 Work out whether these triangles are scalene, right-angled or obtuse-angled.

(a) *ABC*, where $AB = 6$ cm, $BC = 8$ cm and $AC = 9$ cm

(b) *PQR*, where $PQ = 7$ cm, $QR = 4$ cm and $PR = 4.5$ cm

(c) *XYZ*, where $XY = 10$ cm, $YZ = 24$ cm and $XZ = 26$ cm

(d) *DEF*, where $DE = 6.2$ cm, $EF = 4.5$ cm and $DF = 5.4$ cm

(e) *KLM*, where $KL = 12.1$ cm, $LM = 8$ cm and $KM = 4.9$ cm

$a^2 + b^2 = c^2$ right-angled
$a^2 + b^2 > c^2$ scalene
$a^2 + b^2 < c^2$ obtuse

5 Calculate the length of *PQ* in each of these shapes.

(a)

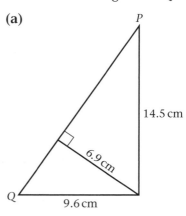
P
14.5 cm
6.9 cm
Q
9.6 cm

(b)

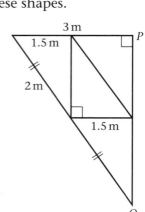
3 m
1.5 m
P
2 m
1.5 m
Q

Exercise 15.3   **Link 15H**

1 Calculate the longest diagonals *AB* of these cuboids.

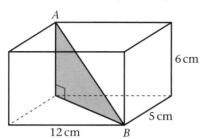

2 A vertical radio mast, 50 metres tall, is situated on a horizontal
piece of land and is secured at points *A*, *B* and *C* by straight
restraining wires. *A*, *B* and *C* are 20, 15 and 30 metres respectively
from the foot of the mast. Calculate the lengths of the three
restraining wires.

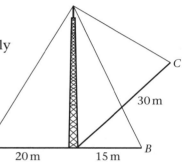

3 The height of this square-based pyramid is
equal to the length of the edges of the
square base. The length of the slant edges
of the pyramid is 10 cm. Calculate the height
of the pyramid.

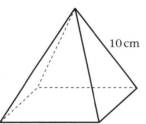

Exercise 15.4   **Link 15I**

1 Write down the equations of these circles.

(a)

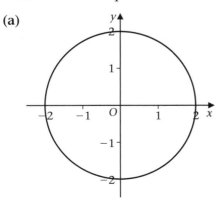

(b)

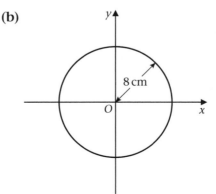

Remember:

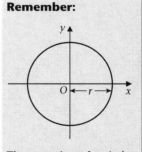

The equation of a circle
with centre at the origin
and radius *r* is
$x^2 + y^2 = r^2$

2   Find the equations of these circles.

   (a)   radius 10 with centre the origin

   (b)   radius 8 with centre the origin

   (c)   radius 1 with centre the origin

   (d)   centre the origin and passes through (2, 2)

   (e)   centre the origin and passes through (6, 8)

16 Basic trigonometry

Exercise 16.1

Links 16A, B

1 Use your calculator to find the value of

(a) sin 30°   (b) cos 60°   (c) tan 45°   (d) sin 56°

(e) tan 78°   (f) cos 54.6°   (g) sin 38.1°   (h) tan 32.5°

> **Remember:** Check that your calculator is working in degree mode.

2 Use your calculator to find the angle a when

(a) $\sin a = 0.866$ (b) $\tan a = 1.192$ (c) $\cos a = 0.5$

(d) $\tan a = 0.603$ (e) $\sin a = 0.621$ (f) $\tan a = 0.6$

(g) $\cos a = 0.75$ (h) $\sin a = 0.5$ (i) $\cos a = (\sqrt{2} \div 2)$

> Press the INV, SHIFT or ARC button to help you find the values of a.

3 Write down the trigometric ratio needed to calculate the side or angle marked x.

(a)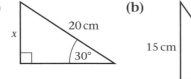

(b)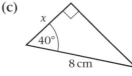

(c)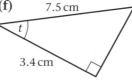

Exercise 16.2

Links 16C, D

1 Calculate the size of the angles marked with a letter.

(a)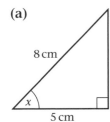

(b)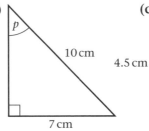

(c)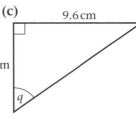

> **Remember:**
>
>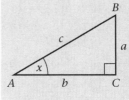
>
> $\sin x = \dfrac{\text{opp}}{\text{hyp}} = \dfrac{a}{c}$
>
> $\cos x = \dfrac{\text{adj}}{\text{hyp}} = \dfrac{b}{c}$
>
> $\tan x = \dfrac{\text{opp}}{\text{adj}} = \dfrac{a}{b}$
>
> You must know these for the exam.

(d) (e) (f)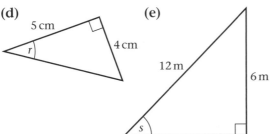

2 Calculate the length of the sides marked with a letter.

(a)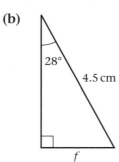

5.6 cm

55°

s

(b)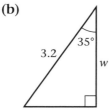

28°

4.5 cm

f

(c)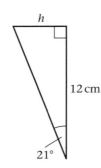

h

12 cm

21°

3 Calculate the length of the sides marked with a letter.
All the lengths are in metres.

(a)

x

12.6

54°

(b)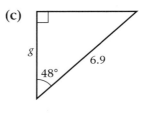

35°

3.2

w

(c)

g

48°

6.9

(d)

6.7

x

45°

(e)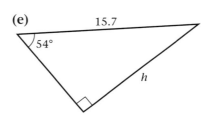

15.7

54°

h

4 Bob places his 5-metre ladder, on horizontal ground, against the
vertical wall of his house. The ladder makes an angle of 65° with the
ground.

(a) How far away from the wall is the foot of the ladder?

(b) How far up the wall does the ladder reach?

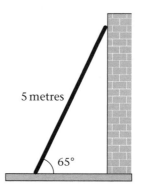

5 metres

65°

Exercise 16.3   **Links 16E, F**

1 Use your calculator to find the value of

(a) sin 120°   **(b)** cos 120°   **(c)** tan 120°   **(d)** sin 210°

(e) cos 210°   **(f)** tan 210°   **(g)** cos 310°   **(h)** tan 310°

(i) sin 310°   **(j)** sin −30°   **(k)** cos −30°   **(l)** tan −30°

(m) sin −300°   **(n)** cos −300°   **(o)** tan −315°   **(p)** sin 180°

Remember:
$\sin(-x) = -\sin x$
$\cos(-x) = \cos x$
$\tan(-x) = -\tan x$

2 Write down the equations for the three graphs in the diagram.

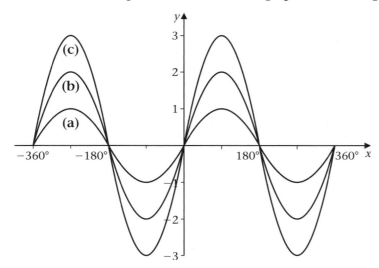

3 Use your calculator to find four values for angle b between $-360°$ and $360°$ in these cases.

Sketch the graphs to help you find all the solutions.

    (a) $\sin b = 0.5$ **(b)** $\cos b = 0.5$ **(c)** $\tan b = 1$

    (d) $\sin b = -0.5$ **(e)** $\cos b = -\frac{1}{3}$ **(f)** $\tan b = -\frac{2}{3}$

    (g) $\sin b = -\frac{1}{4}$ **(h)** $\cos b = 0$

4 On the same grid and for values of x between $-360°$ and $+360°$ draw the graphs of

    (a) $\cos x$ **(b)** $\cos 2x$ **(c)** $\cos 3x$

Label the points where the graph crosses the axes.

5 Solve the following trigonometric equations for values of x between $-360°$ and $+360°$.

    (a) $3 \sin 2x = 2$ **(b)** $2 \cos 3x = 1$ **(c)** $3 \tan 4x = 10$

    (d) $5 \sin 3x = 4$ **(e)** $10 \cos 2x = 3$ **(f)** $2 \tan 5x = 7$

6 Suggest suitable equations for the following graphs.
Check your answers using a graphical calculator.

(a)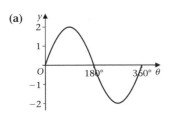

(b)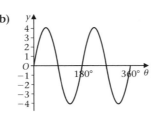

(c)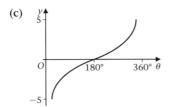

(d)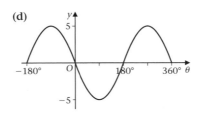

7 This is a graph of the curve $y = \sin x°$, for $0 \leqslant x \leqslant 180$.

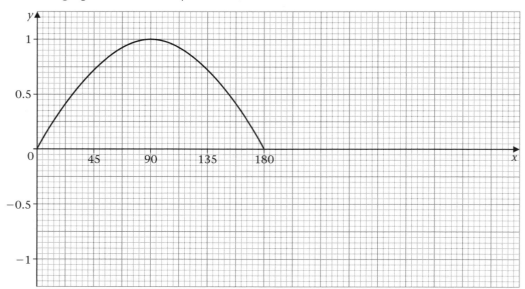

(a) Using the graph, or otherwise, find estimates of the solutions
in the interval $0 \leqslant x \leqslant 360$ of the equations
(i) $\sin x° = 0.2$
(ii) $\sin x° = -0.6$

$\cos x° = \sin(x + 90)°$ for all values of x.
(b) Write down two solutions of the equation $\cos x° = 0.2$.　　　[E]

17 Graphs and equations

Exercise 17.1

Links 17A, B

(a) For questions **1–6** draw graphs with the following equations, taking values of x from -1 to 7.

(b) In each case give the coordinates of the vertex, draw the line of symmetry and write down its equation.

> You may need $\frac{1}{2}$ values to find the minimum point.

1 $y = x^2 - 5x$

2 $y = x^2 - 6x + 1$

3 $y = 2x^2 - 3x + 15$

4 $y = 3x^2 - 16x + 16$

5 $y = (x - 2)^2 + 18$

6 $y = (2 - x)^2 - 3$

In questions **7–12** draw graphs with the following equations, taking values of x from -1 to 5.

7 $y = x^3 - 4x^2 - 8$

8 $y = x^3 - 6x^2 + 6x - 18$

9 $y = x^3 - 3x^2 + 3x - 1$

10 $y = x^3 - 6x^2$

11 $y = -x^3 + 7x^2 - 10x + 12$

12 $y = 8 - 6x + 6x^2 - x^3$

In questions **13–16** draw graphs of the following equations, taking values of x from -3 to 3.

13 $y = x^3 - 12x$

14 $y = x^3 - 3x^2 + 4$

15 $y = 2 + 12x - x^3$

16 $y = 3x^2 - x^3$

Exercise 17.2

Links 17C, D

For questions **1–4** draw graphs with the following equations using values of x from -1 to 5. Draw in the asymptotes and write down their equations.

> **Remember:** You cannot divide by 0. The graph does not touch the asymptotes.

1 $y = \dfrac{10}{x}$

2 $y = 1 - \dfrac{5}{x}$

3 $y = \dfrac{1}{x - 2}$

4 $y = \dfrac{x - 1}{x - 2}$

In questions **5–9** draw the following graphs using the values of x shown in the brackets.

5 $y = x^3 + 3x^2 + \dfrac{1}{x}$ $(-4 \text{ to } 2)$

6 $y = 2x^3 - 5x - 1$ $(-2 \text{ to } 2)$

7 $y = x^2 + \dfrac{1}{x}$ $(-3 \text{ to } 3)$

8 $y = (x - 1)^3 - \dfrac{1}{1 - x}$ (-2 to 4, including 0.9 and 1.1)

9 $y = x^2 + 2x + \dfrac{1}{x}$ (-3 to 3)

Exercise 17.3   **Link 17E**

Solve the following equations correct to 1 d.p. by drawing appropriate graphs. Use the values of x given in the brackets.

> You may need $\frac{1}{2}$ values to complete the graphs.

1 $x^2 - 5x + 3 = 0$ (0 to 5)

2 $2x^2 - 9x + 5 = 0$ (0 to 5)

3 $3x^2 + 2x - 6 = 0$ (-4 to 3)

4 $x^2 + 8x + 14 = 0$ (-7 to -1)

5 $x^3 - x^2 - 3x = 0$ (-3 to 3)   [3 solutions]

6 $x^2 + \dfrac{1}{x} - 4 = 0$ (-3 to 3)   [3 solutions]

7 $x^2 + 2x - \dfrac{1}{x} = 0$ (-3 to 2)   [3 solutions]

8 $x - \dfrac{1}{x} + 1 = 0$ (-3 to 2)   [2 solutions]

Exercise 17.4   **Links 17F, G**

For questions **1–5** find the solution of these equations which lies between the stated limits using trial and improvement. Give your answers to 2 d.p.

1 $x^3 + x = 37$ ($x = 3$ and $x = 4$)

2 $x^3 - x = 20$ ($x = 2$ and $x = 3$)

3 $x^3 + 3x = 15$ ($x = 2$ and $x = 3$)

4 $(x - 7)^3 + x - 13 = 0$ ($x = 8$ and $x = 9$)

5 $2x^3 + 3x = 10$ ($x = 1$ and $x = 2$)

In questions **6–10** solve the equations using trial and improvement. Give your answers to 2 d.p.

6 $x^2 = \dfrac{1}{x} + 1$ **7** $x^2 + \dfrac{3}{x} = 7$

8 $x^2(2x + 1) - 27 = 0$ **9** $x(x - 1)(x - 2) = 15$

10 $(2 - x)x^2 + 40 = 0$

11 A courier firm will take parcels with perimeter of base plus height up to a maximum of 2 metres. If the base is square with side x cm, then the volume is given by $V = x^2 (200 - 4x)$
Use the method of trial and improvement to find the maximum volume that can be sent.

Exercise 17.5

Link 17H

1 The graph shows a short car journey.
  (a) Describe what is happening on the journey between
    (i)   A and B
    (ii) B and C
    (iii) C and D.
  (b) What is the distance travelled in the first 50 seconds?
  (c) How long does it take to travel the first 400 metres?

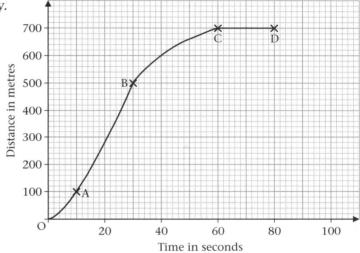

2 The volume of water in a reservoir is measured monthly over a year. The graph shows the measurements for the first nine months.
  (a) Describe briefly how the volume of the reservoir varies between January and August.
  (b) Give possible reasons for the cyclical variation shown in the graph.

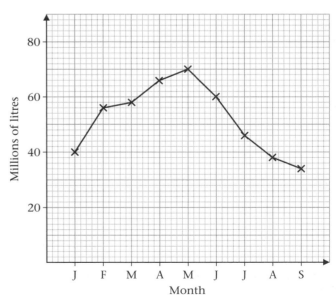

3 The graph shows the height of a skier on a mountain at different times during a certain day.

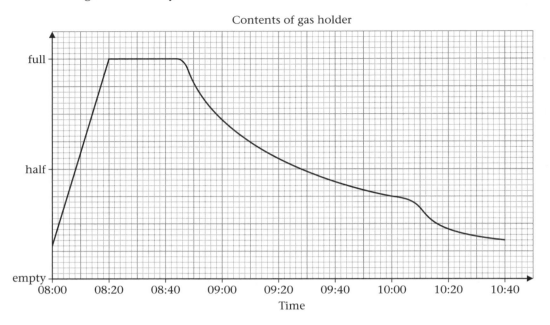

Contents of gas holder

(a) Estimate when the skier is descending quickest.

(b) Describe what is happening between 08:00 and 08:20.

4 The graph shows the number of people in a football ground.

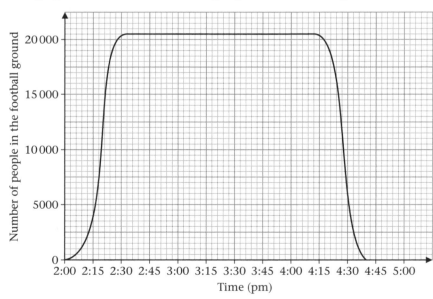

(a) Estimate the time of kick-off.

(b) Estimate the size of the football crowd.

(c) Describe what is happening between 4:15 and 4:45.

5 The graph shows the temperature in an oven.

(a) What temperature was the oven set at?

(b) Give a possible explanation for what happened between 40 and 50 minutes.

(c) When did cooking finish?

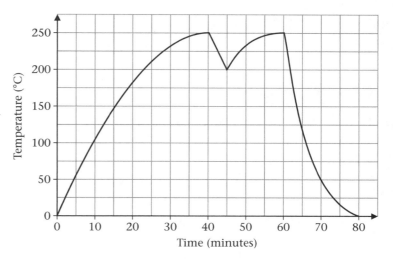

6 Sand flows out of an egg timer in the shape of a cylinder on top of an inverted cone. Sketch a graph to show the depth of sand against time as the sand flows out at a constant rate.

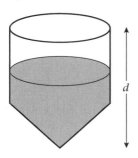

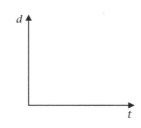

7 Dina accelerates away from traffic lights from rest to a speed of 14 m/s over 6 seconds. She then drives at a constant speed for 20 seconds and then brakes, coming to rest at a junction after a further 7 seconds.
Draw a speed–time graph for Dina's journey.

18 Proportion

1 Which of the following could be examples of direct proportion?
   (a) the number of pens bought and the cost
   (b) the cost of a used car and the age of the car
   (c) the volume of a cube and the length of one edge
   (d) the speed of a car and the distance travelled in one hour
   (e) the time spent on your maths coursework and the final grade
        awarded for it
   (f) the engine size of a car and its fuel consumption

2 $y \propto x$ and $y = 5$ when $x = 12$. Find the value of
   (a) x when $y = 10$
   (b) y when $x = 4$.

   > **Remember:** $y \propto x$ reads
   > as 'y varies directly as x'
   > or 'y is directly
   > proportional to x'.
   > $y = kx$
   > Use the known values of
   > y and x to find k.

3 $y \propto x$ and $y = 12$ when $x = 9$. Find the value of
   (a) x when $y = 72$ (b) y when $x = 3$.

4 The length of a shadow s cast by a tree is directly proportional to
   the height h of the tree. At 2 pm a tree of height 30 metres casts a
   shadow of 18 metres.
   (a) Sketch the graph of the relationship.
   (b) Find the height h of another tree
        when the length of the shadow at 2 pm is 15 metres.
   (c) Find the length s of the shadow at 2 pm of
        a tree that has a height of 40 metres.

5 The amount of electricity w that is used is directly proportional to
   the current used a. When $w = 600$, $a = 2.5$.
   (a) Sketch a graph to show this information.
   (b) Write down a rule that connects w and a.
   (c) When $a = 4$ find the value of w.
   (d) When $w = 1000$ find the value of a.

6 $y \propto x$ and $y = 1.2$ when $x = 0.5$. Find the value, to
   3 significant figures, of
   (a) x when $y = 8.9$ (b) y when $x = 0.3$.

7 p varies directly as q so that $p \propto q$. When $p = 6$, $q = 8$.
   (a) Find the value of p when $q = 24$.
   (b) Find the value of q when $p = 42$.

8 t varies directly as r. When $t = 8.4$, $r = 3.6$.
   (a) Find a rule connecting t and r.
   (b) Find the value of
        (i) t when $r = 4.8$ (ii) r when $t = 9.8$.

9 The circumference C of a circle varies directly as the radius r. When the circumference is 22 cm the radius is 3.5 cm.

(a) Find a rule connecting C and r.

(b) Find the value of
    (i) C when $r = 5.6$ cm   (ii) r when $C = 87$ cm.

10 The volume V of water in a cylinder varies directly as the height h. When the height of the water in the cylinder is 12 cm the volume is 50 ml.

(a) Find a rule connecting V and h.

(b) Find the volume of water when the height is 15 cm.

(c) Find the height of water when the volume is 80 ml.

Exercise 18.2   Link 18E

1 y is directly proportional to the square of x, so that $y = kx^2$. Given that $y = 36$ when $x = 3$

(a) calculate the value of k

(b) calculate the value of y when $x = 5$

(c) calculate the value of x when $y = 64$.

2 p is directly proportional to the cube of q, so that $p = kq^3$. Given that $p = 24$ when $q = 2$

(a) calculate the value of k

(b) calculate the value of p when $q = 3$

(c) calculate the value of q when $p = 375$.

3 A varies in direct proportion to the square of r. Given that when $r = 4$, $A = 50$

(a) calculate the value of A when $r = 3$

(b) calculate the value of r when $A = 100$.

> Find the equation connecting r and A.

4 V varies in direct proportion to the cube of r. Given that when $r = 3$, $V = 113$

(a) calculate the value of V when $r = 5$

(b) calculate the value of r when $V = 40$.

5 The speed S of a car is directly proportional to the square of the braking power P. When the braking power is 35 the speed is 20.

(a) Calculate the value of S when $P = 50$.

(b) Calculate the value of P when $S = 70$.

6 The volume V of a hemisphere varies directly as the cube of the radius r. When the radius is 5 cm the volume is 262 cm^3.

(a) Calculate the volume when the radius is 2 cm.

(b) Calculate the radius when the volume is 50 cm^3.

7 A stone is dropped from the top of a cliff. The distance of the stone from the top varies as the square of the speed. When the speed is 10 metres per second the distance is 5 metres.

    (a) Calculate the distance from the top when the speed is 12 metres per second.

    (b) Calculate the speed when the distance from the top is 30 m.

8 The surface area of a cylinder is directly proportional to the square of the radius. When the area is 12 cm² the radius is 2.5 cm. Calculate the area when the radius is 5 cm.

Exercise 18.3   **Link 18F**

1 y is inversely proportional to x so that $y = \dfrac{k}{x}$.

    When $x = 10$ the value of y is 2.

    (a) Find the value of k.

    (b) Find the value of y when $x = 40$.

    (c) Find the value of x when $y = 10$.

2 p is inversely proportional to q so that $p = \dfrac{k}{q}$.

    When $p = 2$ the value of q is 0.01.

    (a) Find the value of p when $q = 4$.

    (b) Find the value of q when $p = 0.1$.

3 r is inversely proportional to the square of t so that $r = \dfrac{k}{t^2}$.

    When $r = 8$ the value of t is 0.5.

    (a) Find the value of r when $t = 3$.

    (b) Find the value of t when $r = 10$.

4 The gravitational pull F of a planet is inversely proportional to the square of the distance d from the planet. When the distance from the planet is 1000 km the gravitational pull is 12 N.

    (a) Calculate the gravitational pull when the distance is 200 km.

    (b) Calculate the distance when the gravitational pull is 50 N.

> Find the equation connecting F and d.

5 The luminance L of a light source is inversely proportional to the square of the distance d from the light source. When the distance from the light source is 2 metres the luminance is 500 candelas per square metre.

    (a) Calculate the luminace when the distance is 5 metres.

    (b) Calculate the distance when the luminance is 200 candelas per square metre.

6 y is inversely proportional to the square root of x.

    When $y = 5$, $x = 12$.

    (a) Calculate the value of y when $x = 20$.

    (b) Calculate the value of x when $y = 3$.

7 The dispersion factor D of molecules of gas in a spherical container is inversely proportional to the cube of the radius r of the sphere. When the radius is 10 cm the dispersion factor is 4.

   (a) Calculate the dispersion factor D when the radius is 5 cm.

   (b) Calculate the radius r when the dispersion factor is 20.

8 The intensity L of a sound is inversely proportional to the square root of the distance d from the source of the sound. When the distance is 20 metres the sound intensity is 90 W/m^2.

   (a) Calculate the sound intensity L when the distance is 10 metres.

   (b) Calculate the distance when the sound intensity is 120 W/m^2.

19 Quadratic equations

Exercise 19.1 Link 19A

Solve the equations.

1 $(x - 4)(x - 7) = 0$

2 $(x - 2)(x + 3) = 0$

3 $(2x - 1)(3x + 4) = 0$

4 $3(1 - 3x)(3 + 2x) = 0$

5 $(4x - 1)(5x - 2) = 0$

Factorise and solve the quadratic equations.

6 $a^2 - 9a = 0$

7 $a^2 + 2a + 1 = 0$

8 $b^2 - 7b + 6 = 0$

9 $b^2 - 5b = 6$

10 $6c^2 + 11c - 10 = 0$

11 $2x^2 - 6x = 0$

12 $(2x - 1)^2 = 4$

13 $(2 - 3x)^2 = (x + 1)^2$

14 $(x - 3)^2 + 5(x - 3) + 4 = 0$

> Use $(x - 3) = y$.

Exercise 19.2 Links 19B–D

1 Write the following in the form $(x + q)^2 + r$.

    (a) $x^2 + 2x$ **(b)** $x^2 + 5x$ **(c)** $x^2 - 8x$

2 Write the following in the form $p(x + q)^2 + r$.

    (a) $2x^2 + 6x$ **(b)** $3x^2 - 18x + 3$ **(c)** $5x^2 + 40x - 15$

3 Solve these equations. Give your answers correct to 2 d.p.

> Complete the square or use a formula.

    (a) $x^2 + 7x + 9 = 0$ **(b)** $x^2 - 3x - 8 = 0$

    (c) $2x^2 - 11x + 6 = 0$ **(d)** $3x^2 + 8x - 1 = 0$

    (e) $4x^2 - 15x + 5 = 0$ **(f)** $3x^2 = 9 - x$

    (g) $8 = 2x + 11x^2$ **(h)** $7x = 3x^2 - 5$

Exercise 19.3 Links 19E, F

In questions **1–6** find the solutions of the equations, correct to 2 d.p.

> When the denominators have no common factor, the LCM is the product of the denominators.

1 $\dfrac{2}{x + 1} - \dfrac{1}{x + 2} = 1$

2 $\dfrac{4}{2x - 5} - \dfrac{3}{x + 7} = 2$

3 $\dfrac{1}{x} - \dfrac{1}{x - 1} = -3$

4 $\dfrac{2}{(x - 1)(2x - 1)} + \dfrac{3x}{x - 1} = 2$

5 $\dfrac{(x + 1)(3x - 7)}{2x(x - 3)} = 1$

6 $\dfrac{x + 1}{x + 2} = \dfrac{2x + 3}{x + 1}$

7 The area of this square is 40 square units.

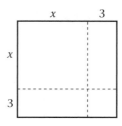

    (a) Form an equation in x.

    (b) Solve your equation.

8 $h = ut - 5t^2$ is a formula which links height in metres (h) with initial velocity in m/s (u) and time in seconds (t).

    (a) If the initial velocity is 25 m/s find the times when the height is 30 metres.

    (b) If the initial velocity is 18 m/s find the times when the height is 16 metres.

    (c) Find the time if the initial velocity is 10 m/s and the height is -15 metres.

9 The diagram is of a right-angled triangle with the lengths of the sides as shown.

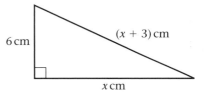

    (a) Use Pythagoras' theorem to form an equation in x.

    (b) Solve the equation to find x.

10 The diagram shows a tent in the shape of a triangular prism. The tent is 10 cm wider than it is high. It is 210 cm long and has a volume of 1386 litres.

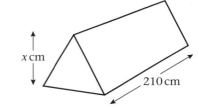

    (a) Work out an expression for the cross-sectional area of the tent in terms of x.

    (b) Use the volume of 1386 litres to form an equation.

    (c) Find the height.

11 The diagram shows a circle with centre O.

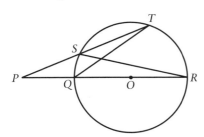

Because triangles PQT and PSR are mathematically similar:

$$PS \times PT = PQ \times PR$$

    (a) Given that $PQ = 5$, $PS = 6$ and $ST = 3$ and the radius is x, form an equation in x and solve it.

    (b) Given that the radius is 9, $PS = 7$ and $ST = 2$, find PQ.

Exercise 19.4 Links 19G, H

1 Solve the simultaneous equations.

(a) $y = 18$
$y = 2x^2$

(b) $y = 4x$
$y = x^2$

(c) $y = 2x - 1$
$y = x^2$

(d) $y = 2x + 7$
$y = x^2 - 4x$

2 Find the coordinates of the points of intersection.

(a) $y + 5x = 12$
$y = 2x^2$

(b) $y = 3x^2 - 2x$
$5y = 3x - 2$

(c) $y = x^2 + 2x$
$7y = 6 - 5x$

(d) $y = x^2 - 1$
$12y = 7x$

> To find the points of intersection, solve the simultaneous equations.

3 In each part solve the simultaneous equations and interpret your solution geometrically.

(a) $x = 2$, $x^2 + y^2 = 16$

(b) $5y = x + 39$, $x^2 + y^2 = 65$

(c) $x^2 + y^2 = 65$, $x = 3y + 5$

> **Remember:** Sketch the graphs.

4 (a) Show that $y = 2x + 5$ is a tangent to the circle $x^2 + y^2 = 5$.

(b) Find the equations of the tangents to the circle $x^2 + y^2 = 5$ that are perpendicular to $y = 2x + 5$.

5 Find the equation of the tangent to the circle $x^2 + y^2 = 50$ which touches the circle at $(1, 7)$.

> The tangent to a circle is perpendicular to the radius.

Exercise 19.5 Link 19I

1 (a) On graph paper draw the graph of $y = 6 + 2x - x^2$ for values of x from -3 to $+4$.

(b) On the same axes draw the graph of $y = \dfrac{7}{x}$ for $-3 \leqslant x \leqslant -\frac{1}{2}$ and $\frac{1}{2} \leqslant x \leqslant 4$.

(c) Use your graphs to find approximate solutions to

(i) $6 + 2x - x^2 = 0$ (ii) $2x - x^2 = -4$ (iii) $6 + 2x - x^2 = \dfrac{7}{x}$

2 (a) On graph paper draw the graph of $y = x^3 - 4x + 4$ for $-3 \leqslant x \leqslant 3$.

(b) On the same axes draw the graph of $y = 2x + 8$.

(c) Use your graphs to find approximate solutions to

(i) $x^3 - 4x + 4 = 4$ (ii) $x^3 - 4x = -8$
(iii) $x^3 - 4x = 2x + 4$ (iv) $2x^3 - 8x + 8 = 3$

3 **(a)** On graph paper draw the graph of $y = 9x - x^3$ for $-3 \leqslant x \leqslant 3$.

(b) On the same axes draw the graph of $y = x - \dfrac{1}{x}$ for

$-3 \leqslant x \leqslant -0.1$ and $0.1 \leqslant x \leqslant 3$.

(c) Use your graphs to find approximate solutions to

(i) $x - \dfrac{1}{x} = 0$

(ii) $x^3 - 9x = 0$

(iii) $x^3 = 9x$

(iv) $9x - x^3 = x - \dfrac{1}{x}$

(v) $\dfrac{1}{x} + 8x - x^3 = 0$

(vi) $\dfrac{1}{x} - x = 2$

20 Presenting and analysing data 2

Exercise 20.1

Links 20A, B

1 A safari park game keeper records the ages of 200 elephants in the park. The results are shown in the table.

Age (years)	0–9	10–19	20–29	30–39	40–49	50–59	60–69	70–79
Frequency	7	11	18	37	46	36	31	14

(a) Calculate an estimate of the mean age.

By drawing a cumulative frequency graph, estimate

(b) the median

(c) the upper quartile

(d) the interquartile range

(e) how many elephants are at least 55 years old

(f) the percentage of elephants that are less than 35 years old.

Plot the cumulative frequencies at the upper class boundaries. Age is rounded down to the nearest whole number. So the upper class boundary of the interval 0–9 is 9.

Remember: The upper quartile is three quarters of the way into the distribution.

2 The table shows the marks obtained by 250 students in a test.

Mark %	Frequency
0–10	1
11–20	3
21–30	6
31–40	12
41–50	47
51–60	62
61–70	79
71–80	25
81–90	13
91–100	2

(a) Draw a cumulative frequency diagram.

(b) What percentage of students scored more than 64 marks?

(c) What percentage of students scored between 32 and 64 marks?

(d) What mark was exceeded by 75% of the students?

Plot the cumulative frequencies at 10, 20, 30, etc.

3 The table shows the age distribution of an island's population.

(a) Draw a cumulative frequency diagram.

(b) Use your diagram to estimate
  (i) the percentage of the population that are 16 years of age or younger
  (ii) the percentage of the population that are between 16 and 65 years of age.

Age	Frequency (10 000)
0–10	38
11–20	67
21–30	102
31–40	87
41–50	48
51–60	25
61–70	12
71–80	7
81–90	4

Exercise 20.2

Links 20C–E

1 The times when students arrive at school are recorded one morning. This frequency table shows the results.

(a) Work out the cumulative frequencies.

(b) Construct the cumulative frequency diagram.

(c) What is the median time of arrival?

(d) Work out the interquartile range.

(e) School starts at 9.00 am. Estimate the percentage of students that are more than 7 minutes early.

(f) The last 5% of students to arrive are punished. Estimate how late these students were.

Time of arrival	Frequency
$t < 8.45$	130
$8.45 \leqslant t < 8.50$	280
$8.50 \leqslant t < 8.55$	520
$8.55 \leqslant t < 9.00$	430
$9.00 \leqslant t < 9.05$	80
$9.05 \leqslant t < 9.10$	30
$9.10 \leqslant t < 9.15$	20
$9.15 \leqslant t$	10

Time is continuous data, so plot the cumulative frequencies at 8.45, 8.50, 8.55, etc.

2 The weekly wages of 100 workers in East Anglia are given in the table.

By drawing a cumulative frequency graph calculate

(a) an estimate of the median

(b) an estimate of the interquartile range

(c) the percentage of workers earning less than £172.

(d) Draw the box plot.

Wage (£)	Frequency
221–240	2
241–260	6
261–280	12
281–300	38
301–320	29
321–340	9
341–360	4

3 The weights of 80 pigs were measured at a farm.

Weight (kg)	Number of pigs
340 to under 345	3
345 to under 350	6
350 to under 355	17
355 to under 360	27
360 to under 365	14
365 to under 370	9
370 to under 375	4

> Draw the box plot below the cumulative frequency diagram.

(a) Calculate an estimate of the mean.

By drawing a cumulative frequency graph, estimate

(b) the median

(c) the upper quartile

(d) the interquartile range

(e) the percentage of pigs that are less than 62 kg.

(f) Draw the box plot.

4 This back-to-back stem and leaf diagram gives information about the time, in hours, for mustard seeds to germinate at different temperatures.

20 °C Time in hours		15 °C Time in hours
	1	5
5, 3, 2	2	1, 4
9, 7, 1, 0	3	2, 6
6, 5, 3, 3, 1	4	2, 2, 4, 4, 6
8, 8, 5, 5, 2, 0	5	3, 6, 7, 7, 8, 8
9	6	1, 1, 1, 2, 3, 5, 9

Key: 0 | 3 | 2 means 30 and 32 hours.

(a) Draw box plots to compare information.

(b) Comment on the germination of the seeds at different temperatures.

Exercise 20.3
Links 20F, G

1 The heights in metres of 30 sunflowers are shown below.

```
1.76   2.45   1.91   0.78   1.90   1.46   2.04   1.87   1.68   2.15
1.26   0.98   2.79   1.86   1.49   1.46   1.76   2.12   2.89   3.20
0.98   2.12   1.56   1.68   1.42   1.34   1.78   2.01   0.46   1.73
```

Remember: In a histogram the areas of the rectangles are proportional to the frequencies they represent.

(a) Draw up a frequency table with class intervals of
$0 < h \leq 1.0$, $1.0 < h \leq 1.5$, $1.5 < h \leq 2.75$, $1.75 < h \leq 2.25$,
$2.25 < h \leq 3.0$, $3.0 < h \leq 3.5$.

(b) Draw a histogram to show this data.

Use frequency density $= \dfrac{\text{frequency}}{\text{class width in standard class intervals}}$

2 The time to cycle 1000 m was measured for members of a cycle club. The times taken were recorded in a table.

Draw a histogram to show this data.

Time (t) to cycle 1000 m (s)	Frequency
$120 < t \leq 150$	6
$150 < t \leq 170$	15
$170 < t \leq 210$	18
$210 < t \leq 300$	9
$250 < t \leq 300$	2

3 This histogram represents the lengths of cuttings planted by a gardener one weekend.

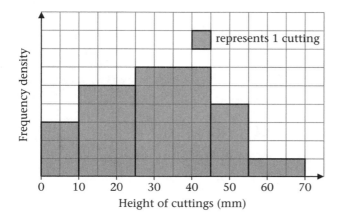

represents 1 cutting

(a) Write down the number of cuttings with length less than 25 mm.

(b) Write down the number of cuttings with length greater than or equal to 10 mm and less than 55 mm.

(c) What was the total number of cuttings in the sample?

4 The diagrams represent the histograms for three distributions. In each case, sketch the cumulative frequency curve.

(a)
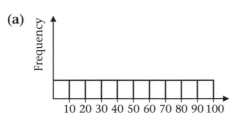

(b)
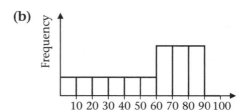

(c)
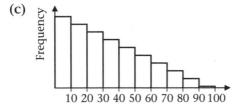

5 The local education authority wants to compare the LEA test results in two schools. They gathered the following information.

	School A	School B
Maximum %	15	30
Minimum %	98	96
Median	58	72
Upper quartile %	76	84
Lower quartile %	28	40

(a) Choose an appropriate scale and draw the box and whisker diagram for each school.

(b) Which school do you believe had the better results? Give your reasons.

> **Remember:** When you compare the information in two box plots, always draw the diagrams lined up, one above the other.

6 This histogram represents the number of spectators at professional rugby matches on one Saturday in 1998.

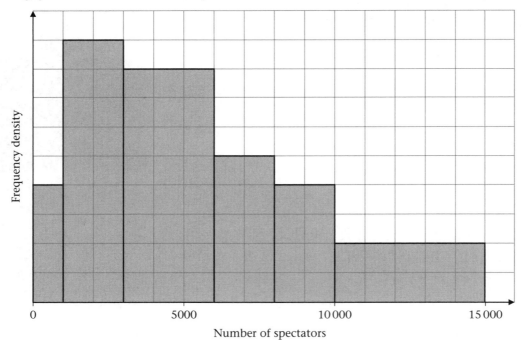

Number of spectators

No match had more than 15 000 spectators.
At five matches the number of spectators was greater than or equal to 6000 and less than 8000.

(a) Use the above information to complete this frequency table.

(b) Calculate the total number of professional rugby matches played on that Saturday.

Number of spectators (n)	Frequency
$0 \leqslant n < 1000$	
$1000 \leqslant n < 3000$	
$3000 \leqslant n < 6000$	
$6000 \leqslant n < 8000$	5
$8000 \leqslant n < 10\,000$	
$10\,000 \leqslant n < 15\,000$	

21 Advanced trigonometry

Exercise 21.1 Link 21A

1 Calculate the area of each of the triangles to 2 d.p.

Remember:
Area $= \frac{1}{2}ab \sin C$

(a)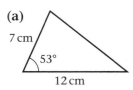

7 cm, 53°, 12 cm

(b)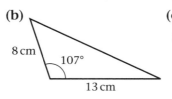

8 cm, 107°, 13 cm

(c)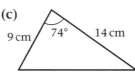

9 cm, 74°, 14 cm

(d)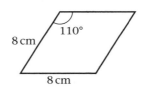

10 cm, 10 cm, 10 cm

(e)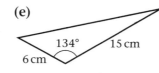

134°, 15 cm, 6 cm

(f)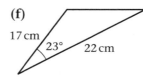

17 cm, 23°, 22 cm

2 Calculate the area of the rhombus.

8 cm, 110°, 8 cm

Split the rhombus into two triangles.

3 A builder fences off a triangular building plot XYZ. $XY = 42$ metres, $XZ = 35$ metres and the angle at $X = 98°$. Calculate the area of the building plot.

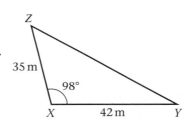

Z, 35 m, 98°, X, 42 m, Y

Remember:
$\sin(180° - x) = \sin x$

4 A triangle PQR has an area of 80 cm². The angle RPQ is acute. $PQ = 24$ cm, $PR = 12$ cm. Calculate the size of the angle RPQ. Give your answer correct to the nearest degree.

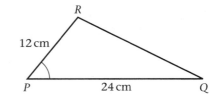

R, 12 cm, P, 24 cm, Q

5 A triangle ABC has an area of 483 cm². The angle at B is obtuse. $BA = 47$ cm, $BC = 32$ cm. Calculate the size of the angle ABC. Give your answer correct to the nearest degree.

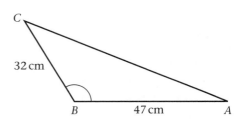

C, 32 cm, B, 47 cm, A

6 A triangle STU has an area of 30 cm². $ST = 15$ cm, $SU = 8$ cm. Calculate both of the possible sizes of the angle TSU.

Exercise 21.2

1 Work out the sides or angles marked with a letter.

(a)

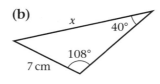

9 cm, x, 37°, 42°

Remember: Use the sine rule.

(b)

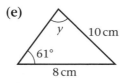

x, 40°, 108°, 7 cm

(c)

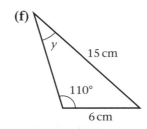

80°, x, 53°, 7 cm

(d)

8 cm, 12 cm, 41°, y

(e)

y, 10 cm, 61°, 8 cm

(f)

y, 15 cm, 110°, 6 cm

2 A lighthouse, *L*, lies 52 km due north of a marker buoy, *B*.
A trawler, *T*, lies on a bearing of 035° from *B* and on a bearing of 048° from *L*. Work out the distance from *L* to *T*.

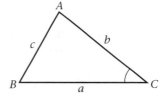

Draw a sketch first.

3 The area of a triangle *ABC* can be written as $\frac{1}{2}ab \sin C$.

(a) Write the area in two other ways, using angles *A* and *B* respectively.

(b) By equating the expressions for the area of *ABC*, establish the **sine rule**.

A, *c*, *b*, *B*, *a*, *C*

4 *PQR* is a triangular plot of land.
PQ = 32 m, *PR* = 37 m and the angle *PQR* = 45°.
Work out the angle *PRQ*.
Give your answer to the nearest degree.

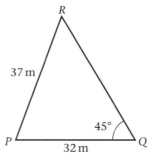

R, 37 m, 45°, *P*, 32 m, *Q*

5 The diagram shows the relative positions of three villages, Ashwell (*A*), Bredbury (*B*) and Crimpton (*C*).
The bearing of Bredbury from Ashwell is 023°.
The bearing of Crimpton from Ashwell is 075°.
The bearing of Crimpton from Bredbury is 114°.

(a) Show that the angle *ACB* = 39°.

The distance from Ashwell to Bredbury is 8.6 km.

(b) Work out the distance from Bredbury to Crimpton.

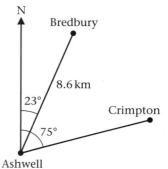

N, Bredbury, 8.6 km, 23°, Crimpton, 75°, Ashwell

6 A woman leaves her home, *H*, and walks 1.2 km due east to reach a barn, *B*. At *B*, she turns through an angle of 130° and continues to walk in a straight line until she reaches a shed, *S*.
The angle *BSH* is 27°.
Calculate how far the woman would have walked if she had walked in a straight line from her home to the shed.

> Draw a sketch first.

7 In a triangle *XYZ*, *XY* = 12 cm, *YZ* = 7 cm and the angle at *X* = 23°.

    (a) Show that there are two possible values for the angle at *Z*.

    (b) Calculate each of the values of the angle at *Z*.

Exercise 21.3

                                                       Link 21E

1 Calculate each of the sides or angles marked with a letter.

> **Remember:** Use the cosine rule.

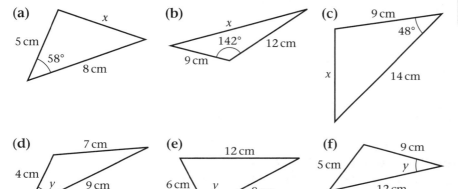

2 A ship leaves a port, *P*, and travels 46 km due north to reach a lighthouse, *L*. At *L* the ship turns on to a bearing of 310° and travels a further 71 km to reach a marker buoy, *B*. At *B* the ship turns again and travels in a straight line back to *P*.
Calculate the total distance travelled by the ship.

> Draw a sketch first.

3 The lengths of the sides of a triangle are 5 cm, 12 cm and 15 cm.
Work out the three angles of this triangle.
Give your answers correct to the nearest degree.

4 In a triangle *ABC*, the lengths of the sides are
    AB = *c*, *BC* = *a* and *AC* = *b*
The angle *ACB* = 120°.
Show that
    $c^2 = a^2 + b^2 + ab$

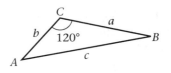

5 *PQRS* is a parallelogram with *PQ* = 12 cm, *QR* = 7 cm and the angle *PQR* = 150°. Work out the lengths of the two diagonals of the parallelogram.

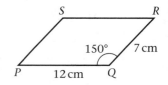

6 In triangle ABC, $AB = 5$ cm, $AC = x$ cm,
$BC = 2x$ cm and angle $BAC = 60°$.
Show that $3x^2 + 5x - 25 = 0$

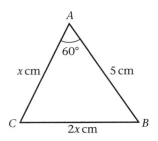

7 The perimeter of a triangle is 50 cm.
The length of one of the sides is 24 cm, the length of
another of the sides is 23 cm. Calculate, to the nearest
degree, the size of the smallest angle of this triangle.

8 A port, P, is 23 km due north of a harbour, H.
At 12:00 a yacht sets out from P and travels at 12 km/h
on a bearing of 072°.
Also at 12:00, a ship sets out from H and travels on a
straight course in a general north-easterly direction.
The yacht and the ship meet up at 14:00 hours.
Calculate

(a) the speed of the ship

(b) the bearing on which the ship travelled.

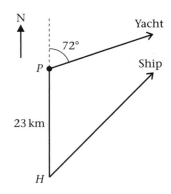

Exercise 21.4   Link 21F

1 $VABC$ is a solid triangle-based pyramid.
The vertex V is vertically above B with $VB = 24$ cm.
The horizontal base is a triangle ABC with the angle at B being 90°.
$AB = 10$ cm   and $BC = 7.5$ cm.

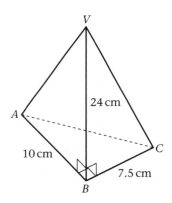

(a) Calculate the lengths of
    (i) AC **(ii)** VC **(iii)** VA

(b) Calculate the angles
    (i) BVC **(ii)** VAB **(iii)** AVC

(c) Calculate the surface area of $ABCV$.

2 $ABCDEF$ is a wedge.
The rectangular base, $BCDE$, is horizontal with
$BC = 8$ cm and $CD = 15$ cm.
The three faces ABC, FED and $ABEF$ are all vertical.
$ABEF$ is a rectangle with $AB = 6$ cm and $AF = 15$ cm.
ABC and FED are triangles with right angles at
B and E respectively.

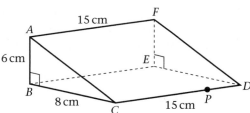

(a) Calculate the lengths of
    (i) AC **(ii)** BD **(iii)** FC

(b) Calculate the angles
    (i) ADC **(ii)** FCE

A point P lies on CD such that $CP = 12$ cm.

(c) Calculate the angles
    (i) FPE **(ii)** EPB

> Sometimes it helps to
> draw separately the
> triangle you are using.

3 *VABCD* is a rectangular-based pyramid.
The vertex *V* is vertically above *M*, the midpoint of the
horizontal base *ABCD*.
AB = 16 cm, *AC* = 34 cm and *VA* = 25 cm.

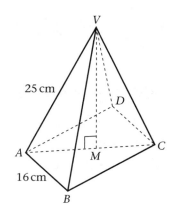

(a) Calculate the lengths of
  (i) *BC*   (ii) *VM*

(b) Calculate the angles
  (i) *VBM*   (ii) *BVA*   (iii) *AVD*

4 *PQRS* is a regular tetrahedron with
each side of length 10 cm.
It is placed so that the face *PQR* is
on a horizontal table.
Work out the height of *S* above
the table.

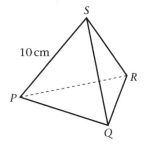

5 The diagram shows a tetrahedron *VABC*.
The horizontal base is a triangle *ABC* with
AB = 16 cm, *BC* = 24 cm and the angle *ABC* = 120°.

The vertex *V* is vertically above *B* and *VB* = 30 cm.

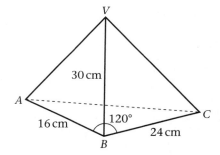

(a) Calculate the area of the base *ABC*.

(b) Calculate the lengths of
  (i) *VA*   (ii) *VC*   (iii) *AC*

(c) Calculate the angles
  (i) *AVB*   (ii) *VCB*   (iii) *AVC*
  (iv) *ACV*   (v) *BAC*

22 Advanced mensuration

Exercise 22.1

In this exercise, give your answers to 3 significant figures where appropriate.

1 Calculate the arc length and the area of the sectors of these circles.

Remember:

Arc length $= \dfrac{\theta}{360}$ of the circumference

Area of sector $= \dfrac{\theta}{360}$ of the area

(a)

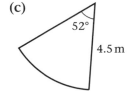

(b)

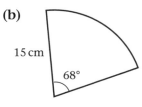

(c)

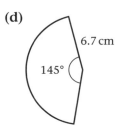

(d)

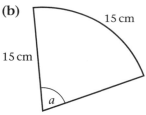

2 Calculate the size of the angles marked a in these sectors of circles.

(a)

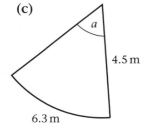

(b)

(c)

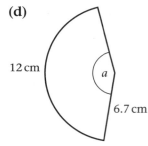

(d)

3 The area of a sector of a circle of radius 5 cm is 10 cm². Calculate the angle at the centre of the sector.

4 The arc length of the sector of a circle is 30 cm when the angle at the centre is 150°. Calculate the radius of the circle.

5   Calculate the shaded areas in these sectors.

Area of sector =
area of segment − area of triangle

(a)

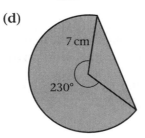

15 cm

72°

(b)

4.5 m

52°

(c)

20 cm

240°

(d)

7 cm

230°

6   A shot-put competition takes
place in a circle of diameter
2 metres and a landing area
that is the sector of a circle
with radius 20 metres and an
angle of 38°. The centre of
both circles is the same.
Calculate the total area of the sports field taken up by this shape.

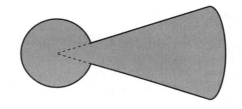

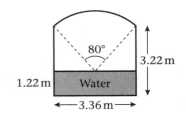

7   The entrance to a tunnel on a canal is in the shape of a
rectangle and the sector of a circle. The width of the rectangle
is 3.36 metres and the height is 3.22 metres. The centre of the
arc of the circle is on the water line of the canal and has an
angle of 80°. The depth of the water in the canal is 1.22 metres.
Calculate the perimeter of the tunnel entrance that is above
the water line.

80°

3.22 m

1.22 m   Water

←—3.36 m—→

8   A stained glass window in a church is made up of sectors
of circles. The outer circle has a radius of 2 metres. Each
inner arc is centred on the outer circle, and has radius √8 m
and an angle of 90° at its centre.

The circumference and all the inner arcs represent the lead
beading that holds the coloured glass in place.
Calculate the length of lead beading used in this window.

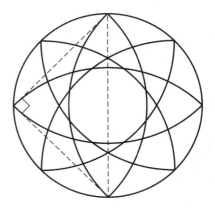

Exercise 22.2

In this exercise, give your answers to 2 decimal places where appropriate.

1 A cylindrical tin of fruit has a base radius of 3 cm and a height of 12 cm.

   (a) Calculate the volume of the tin of fruit.

   (b) Calculate the surface area of the tin.

> **Remember:** The surface area is the total of the areas of the separate surfaces of a solid.

2 A chocolate bar of length 15 cm is in the shape of a triangular prism. The triangular ends are equilateral triangles with edges of 5 cm.

   (a) Calculate the volume of the chocolate bar.

   (b) Calculate the surface area of the bar.

> Find the height of the triangle.

3 A hexagonal steel bar has a length of 30 cm. The length of each edge of the hexagonal end is 0.5 cm. Calculate

   (a) the volume of the bar   **(b)** the surface area of the bar.

4 A cylindrical waste bin has a volume of 40 litres. The radius of the base is 15 cm. Calculate the height of the bin.

> **Remember:**
> 1 litre = 1000 cm^3

5 Calculate the volume of a cone that has a base radius of 10 cm and a height of 15 cm.

> **Remember:** Volume of a cone $= \frac{1}{3}\pi r^2 h$

6 A square-based pyramid has a height of 20 cm. The length of each edge of the square base is 10 cm. Calculate the volume of the pyramid in litres.

> **Remember:** Volume of a pyramid $= \frac{1}{3} \times$ area of base $\times$ height

7 A solid cube of steel has a volume of 1 litre. The cube is melted down and recast as a cylinder of radius 5 cm. Work out the length of the cylinder.

8 A cone has a volume of 500 cm^3 and a height of 12 cm. Calculate the radius of the base.

9 Calculate the volume and surface area of a sphere of

   (a) radius 5 cm   **(b)** radius 0.5 cm   **(c)** radius 1 m

> **Remember:**
> Volume of sphere $= \dfrac{4\pi r^3}{3}$
> Surface area of sphere $= 4\pi r^2$

10 A hexagonal steel bar has a volume of 12 cm^3. The length of the bar is 10 cm. Calculate the length of one edge of the hexagonal end.

11 Calculate the volume of a cube that has a surface area of 12 cm^2.

12 Calculate the volume of a sphere that has a surface area of 1000 cm^2.

Exercise 22.3 Link 22H

In this exercise, give your answers to 3 s.f. where appropriate.

1 Terry makes a model of a World War II Spitfire. He uses a scale of 1 : 100. The wingspan of the real plane is 11 metres.

   (a) Calculate the wingspan of the model plane.

   The area of one of the circular markings on the real plane is 600 cm².

   (b) Calculate the area of one of the circular markings on the model plane.

   The volume of a fuel tank in the real plane was 400 litres.

   (c) Calculate the volume of a fuel tank in the model plane.

> **Remember:** If the scale factor (length) is k, the scale factor (area) is k^2, and the scale factor (volume) is k^3.

2 A manufacturer of wheelbarrows makes them in 3 sizes: small, medium and large. The wheelbarrows are similar in shape. The ratio of the lengths of the wheelbarrows is 2 : 3 : 4.

   (a) If the area of the top of the medium wheelbarrow is 1 m² calculate the areas of the tops of the other wheelbarrows.

   (b) The small wheelbarrow will hold 0.6 m³. How much do the other two wheelbarrows hold?

   The radius of the wheel of the medium wheelbarrow is 10 cm. The volume of the wheel of the large wheelbarrow is 3 litres.

   (c) Calculate the area of cross-section of the wheels of all three wheelbarrows.

   (d) Calculate the volumes of the wheels of the medium and small wheelbarrows.

3 The manufacturers of bottles of a soft drink make it in 1-litre, 2-litre and 3-litre bottles. All the bottles are similar in shape.

   (a) Calculate the ratio of the heights of the three bottles.

   (b) Calculate the ratio of the surface areas of the three bottles.

4 Two similar bottles of 'Sunshine Cola' have lengths that are 12 cm and 15 cm.

   (a) The volume of the large bottle is 1.5 litres. Calculate the volume of the smaller bottle.

   (b) The area of the label on the small bottle is 12 cm². Calculate the area on the label of the large bottle.

5 Mr Green sells ice cream cones that have similar shapes. The diameter of the small cone is 5 cm and the diameter of the large cone is 8 cm. Each cone is filled with ice cream.

   (a) If the volume of ice cream in the small cone is 100 cm³ calculate the volume of ice cream in the large cone.

   (b) If the surface area of the large cone is 268 cm² what is the surface area of the small cone?

6 A scale model of a ship is built on a scale of 1 : 200. Copy this table into your book and complete the information.

Measurement	Ship	Model of ship
Height of mast	30 m	. . . . . . cm
Area of deck		9.5 cm²
Volume of hold	15 000 m³	
Number of portholes	2000	

7 The height of the whole cone is 15 cm. An identical cone has a small cone at the top removed, leaving the bottom section (frustum). The height of this bottom section is 10 cm. The surface area of the small cone is 30 cm². The volume of the large cone is 250 cm³.

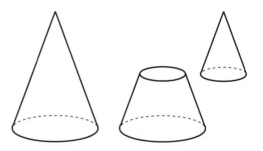

(a) What is the scale factor of the lengths of the large cone to the small cone?

(b) Work out the surface area of the large cone.

(c) Calculate the volume of the frustum of the cone.

Exercise 22.4

Link 22I

1 A toppling clown is made from a hemisphere and a cone.

(a) Calculate the volume of the shape.

(b) Calculate the surface area of the shape.

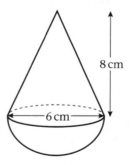

Curved surface area of cone
= $\pi r l$
where l is slant height.

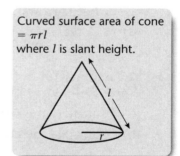

2 A trumpet mute is made in the shape of a truncated (cut-off) cone.

(a) Calculate the volume of the trumpet mute.

(b) Calculate the surface area of the trumpet mute.

3 A model rocket is made from a solid cylinder and a cone. The radius of the cylinder and the cone is 5 cm. The cylinder has height 20 cm and the cone has height 12 cm.

(a) Calculate the volume of the shape.

(b) Calculate the surface area of the shape.

23 Exploring numbers 2

Links 23A, B

Exercise 23.1

1 Convert the following fractions into decimals and indicate which of the decimals are terminating and which are recurring.

$$\frac{2}{3}, \quad \frac{1}{4}, \quad \frac{4}{5}, \quad \frac{5}{6}, \quad \frac{9}{13}$$

2 Write out the recurring decimals which are equivalent to the fractions.

$$\frac{1}{9}, \quad \frac{2}{9}, \quad \frac{3}{9}, \quad \frac{4}{9}, \quad \frac{5}{9}, \quad \frac{6}{9}, \quad \frac{7}{9}, \quad \frac{8}{9}$$

Explain the pattern of recurring decimals.

3 Find the decimals equivalent to the fractions.

$$\frac{1}{15}, \quad \frac{2}{15}, \quad \frac{3}{15}, \quad \cdots \quad \frac{12}{15}, \quad \frac{13}{15}, \quad \frac{14}{15}$$

(a) Separate the decimals into two distinct sets.

(b) Explain the relationship between the two sets.

4 Find fractions which are equivalent to the recurring decimals.

(a) $0.3333\ldots$ **(b)** $0.8888\ldots$ **(c)** $0.575757\ldots$

(d) $0.01230123\ldots$ **(e)** $5.232323\ldots$ **(f)** $3.8565656\ldots$

Remember: Let the fraction $= x$, then multiply it by 10^n, where n is the number of decimal places in the recurring pattern.

5 Find the fraction which is equivalent to the recurring decimal $0.0999999\ldots$

Exercise 23.2

Link 23C

1 Solve $x^2 = 40$, leaving your answer in its most simplified surd form.

2 The area of a square is $90\,\text{cm}^2$. Find the length of one side of the square. Give your answer in its most simplified surd form.

3 Simplify

(a) $\dfrac{1}{\sqrt{5}}$ **(b)** $\dfrac{2}{\sqrt{7}}$ **(c)** $\dfrac{\sqrt{6}}{\sqrt{7}}$

(d) $\dfrac{1}{\sqrt{13}}$ **(e)** $\dfrac{\sqrt{128}}{\sqrt{2}}$ **(f)** $\dfrac{\sqrt{11}}{\sqrt{22}}$

4 A rectangle has sides of length $5 + \sqrt{3}$ and $5 - \sqrt{3}$ units. Work out, in their most simplified form

(a) the perimeter of the rectangle

(b) the area of the rectangle

(c) the length of a diagonal of the rectangle.

5 Solve the equations.

(a) $x^2 - 10x - 32 = 0$ (b) $x^2 + 6x + 3 = 0$

6 Show that: $\dfrac{1}{3\sqrt{13}} = \dfrac{\sqrt{13}}{39}$.

7 Rationalise the denominators.

(a) $\dfrac{1}{2 + \sqrt{2}}$ (b) $\dfrac{\sqrt{2}}{5 - \sqrt{2}}$

(c) $\dfrac{33}{\sqrt{3} + 4}$ (d) $\dfrac{2\sqrt{2}}{\sqrt{2} + \sqrt{8}}$

> Rationalise the denominator of $\dfrac{\sqrt{2}}{5 + \sqrt{2}}$.
>
> Multiply top and bottom by $5 - \sqrt{2}$ because
>
> $(5 + \sqrt{2})(5 - \sqrt{2})$
>
> $\quad = 5 \times 5 - 5\sqrt{2} + 5\sqrt{2} - \sqrt{2}\sqrt{2}$
>
> $\quad = 25 - 2 = 23$
>
> $\dfrac{\sqrt{2}(5 - \sqrt{2})}{(5 + \sqrt{2})(5 - \sqrt{2})} = \dfrac{5\sqrt{2} - 2}{23}$

Exercise 23.3   **Links 23D–F**

1 Write the lower bound and upper bound for these measurements, to the given degree of accuracy.

(a) To the nearest unit.

        5,   15,   105,   115,   1005

(b) To the nearest 1000.

        15 000,   1000,   9000,   5000,   10 000

(c) To the nearest 0.5 unit.

        3.5,   11,   17.5,   21,   22.5

(d) To the nearest 0.25 unit.

        11,   3.25,   9.75,   50,   17.5

2 The 400 m record at a running track in Stevenage is 44.152 seconds correct to the nearest $\frac{1}{1000}$ of a second.

(a) Write down the upper bound and lower bound for this record.

(b) A runner records a lap time of 44.1516 seconds. Explain whether this is a new record.

3 Using the upper bound and lower bound for these measurements, find the upper and lower bounds for these quantities. The degree of accuracy of each measurement is given. Use the π button on your calculator when needed.

(a) The area of a rectangle with sides 5 cm and 9 cm, both measured to the nearest cm.

(b) The perimeter of a rectangle with sides 5 cm and 9 cm, both measured to the nearest cm.

(c) The area of a circle with radius 4.2 cm, measured to 1 d.p.

(d) The circumference of a circle with radius 5.19 cm, measured to 2 d.p.

(e) The area of a triangle with base length 6.1 cm and perpendicular height 12.5 cm, both measured to the nearest mm.

(f) The area of a triangle in which two sides are of length 8 cm and 12 cm and the size of the angle between them is 62°, all measured to the nearest unit.

(g) The volume of a cone with
   (i) radius 25 cm and height 40 cm, both measured to the nearest 5 cm
   (ii) radius 25 cm and height 40 cm, both measured to 2 s.f.
   (iii) radius 25.0 cm and height 40.0 cm, both measured to 1 d.p.

> **Remember:**
> When **adding** two measurements $a + b$
>   Add upper bounds to get the maximum value
>   Add lower bounds to get the minimum value
> When **subtracting** two measurements $a - b$
>   Subtract lower bound b from upper bound a to get the maximum value
>   Subtract upper bound b from lower bound a to get the minimum value
> When **multiplying** two measurements $a \times b$
>   Multiply upper bounds to get the maximum value
>   Multiply lower bounds to get the minimum value
> When **dividing** two measurements $a \div b$
>   Divide upper bound a by lower bound b to get the maximum value
>   Divide lower bound a by upper bound b to get the minimum value

4 Calculate the upper bound and lower bound for the difference in length of two pieces of string. One measures 160 cm, the other 82 cm, both measured to 2 s.f.

5 A car travels 90 km in 1.5 hours, where both measurements are correct to 2 s.f. Calculate the upper bound and lower bound of the car's average speed in km/h.

6 The diagram shows triangle ABC where angle $ACB = 105°$ to 3 s.f., $AC = 18$ cm to 2 s.f. and $BC = 25$ cm to 2 s.f. Use the cosine rule to calculate the lower bound and the upper bound of the length AB.

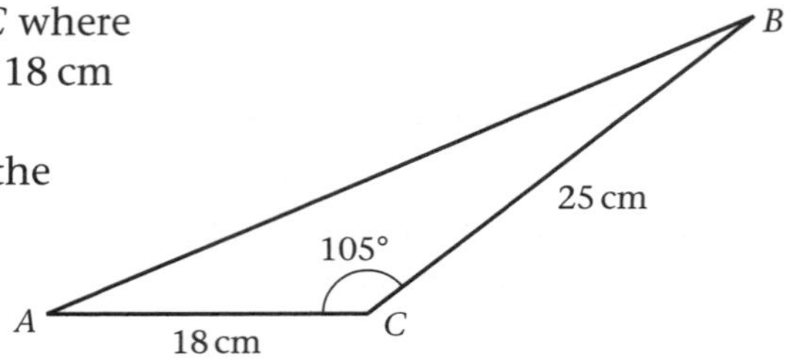

7 A sphere has a radius of 0.45 cm correct to 2 d.p. Calculate the upper bound and lower bound of the volume and surface area of the sphere.

8 $x = 20$ and $y = 30$, correct to the nearest 5.
Work out the minimum possible value of

$$\frac{x + y}{x}$$

9 $x = 4$ and $y = 0.07$, correct to 1 significant figure.

$$z = y - \frac{x - 8}{x}$$

Work out the least upper bound for z.

24 Probability

1 Twelve equal-sized balls are labelled from
1 to 12.
The balls are placed in a bag.
One of the balls is selected at random.
Work out the probability that the selected
ball will be labelled

> **Remember:**
> $$P(event) = \frac{\text{the number of ways the event can occur}}{\text{the total number of possibilities}}$$

(a) 3   (b) a multiple of 3

(c) a prime number   (d) an even number

(e) an odd number   (f) a triangular number

(g) a square number.

2 Fifteen chocolates are put into a bag.
8 of the chocolates are milk, 4 of the chocolates are plain and the
remainder are white.
Suzanne selects a chocolate from the bag at random.
Calculate the probability that this chocolate will be

> **Remember:**
> $P(\text{not } A) = 1 - P(A)$
> $P(A \text{ or } B) = P(A) + P(B)$

(a) milk   (b) not milk

(c) not white   (d) either white or milk.

3 A selection pack contains 36 packets of crisps. In the pack there
are

> 12 bags of plain crisps
> 8 bags of cheese and onion crisps,
> 7 bags of salt and vinegar crisps
> 5 bags of smoky bacon crisps
> 4 bags of prawn cocktail crisps.

On a dark night Savita selects a packet of crisps from the pack at
random.
Write down the probability that she will select

(a) a bag of plain crisps

(b) a bag of crisps that are not plain

(c) a bag of cheese and onion crisps

(d) a bag of smoky bacon crisps

(e) a bag of crisps which is either salt and vinegar or prawn
cocktail

(f) a bag of crisps which is neither plain nor cheese and onion.

4 Kevin has a spinner in the shape of a regular pentagon, and a normal dice.
The five sections of the spinner are labelled 1, 2, 3, 4, 5.
Kevin spins the spinner once and rolls the dice once.
He records the outcome, writing the number shown on the spinner first.
For example, 5 on the spinner and 3 on the dice is recorded as (5, 3).

  (a) Write down the full list of all the possible outcomes.

  (b) Find the probability of these outcomes.

> You can use a sample space diagram to show all the possibilities. The total number of possibilities = 30.

      (i)   (2, 4)
      (ii) (3, 7)
      (iii) the two numbers add up to 10
      (iv) both of the numbers are prime
      (v) either one of the numbers is 3
      (vi) the difference between the two numbers is 1
      (vii) the second number is double the first number.

5 Three coins are tossed simultaneously.

  (a) Using H and T for heads and tails, list all of the 8 possible outcomes.

  (b) Work out the probability that
      (i)   all three coins land heads
      (ii) two coins land tails
      (iii) at least one coin lands tails
      (iv) at least two of the coins land heads.

> 'at least one' means one or more.

6 A 'singles only' dartboard has 20 equal sectors marked from 1 to 20.
It has no bullseye and no spaces for doubles or trebles.
An ordinary dice is a cube with its six faces marked from 1 to 6.
When Marco throws a dart at the board he will hit one of the numbers at random.

Marco rolls the dice and throws a dart at the board.
The number on the upper face of the dice will be used as the
x-coordinate and the score on the dartboard will be used as
the y-coordinate of a point on a grid. Calculate the probability
that when the dice is rolled and the dart is thrown the point
generated on the grid will lie on the line

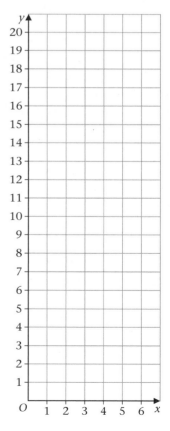

  (a) $x = 4$ **(b)** $y = 9$
  (c) $y = 2x$ **(d)** $x = y$
  (e) $y = 3x$ **(f)** $x + y = 10$
  (g) $y = 2x + 1$ **(h)** $y = 20 - x$
  (i) $x - y = 3$ **(j)** $x + 3y = 8$

Exercise 24.2   Link 24C

1 Here are the final scores from 100 football matches.

2–1	1–1	0–0	3–1	1–3	2–0	2–0	2–1	6–0	3–1
3–0	5–1	0–0	2–1	0–3	1–1	1–2	0–7	4–2	5–0
2–0	3–0	2–1	4–1	2–0	5–1	1–2	1–3	3–2	0–3
1–2	1–2	0–0	0–0	0–1	2–3	1–0	4–2	1–0	1–1
0–0	1–2	1–3	2–0	1–1	1–0	1–1	2–0	0–0	0–1
1–1	3–1	2–2	1–1	0–1	2–1	1–4	2–0	3–1	5–0
1–2	1–1	1–1	1–1	0–2	0–1	4–1	3–2	2–1	1–2
2–0	2–1	1–0	1–1	1–1	3–1	0–0	1–3	1–2	2–1
4–2	3–0	0–1	1–0	2–2	1–1	2–1	1–1	0–1	0–0
6–1	4–0	0–0	6–1	1–1	1–1	3–1	2–0	0–0	1–1

The first score is the number of goals scored by the home team.
The second score is the number of goals scored by the away team.

A football match is chosen at random.

(a) Using this evidence, and this evidence alone, make estimates
of the probability that
(i) the game will end as a draw
(ii) there will be a total of 3 goals in the match
(iii) there will only be a one-goal difference
between the two scores.

(b) Next month there will be exactly 250 football
matches played. Using only the evidence from
the data you have been provided with, give the
best estimate for the number of these matches
that will finish as a draw.

> **Remember:**
> Relative frequency
>
> $= \dfrac{\text{number of times event occurs}}{\text{total number of trials}}$
>
> Relative frequency can be used as
> an estimate of the probability.

2 Shortly before an election a survey was conducted to
find out information about people's voting intentions.
1300 people were asked which of the four candidates
they intended to vote for. The results of the survey are
in the table.

Candidate	Number who intend to vote
Adey	250
Burrows	450
Cresson	372
Davis	228

(a) On the day of the election a person was stopped
in the street at random as they were going to vote.
Work out, giving your reasons, the best estimate of
the probability that this person intended to vote for
(i) Burrows   **(ii)** Davis.

(b) In the actual election 16 500 votes were cast. Give, with
reasons,
the best estimate of the number of votes cast for
(i) Adey   **(ii)** Cresson.

Exercise 24.3   **Links 24D, E**

1 Next Monday, Steve and Linda are both taking their driving test.
The probability of Steve passing is 0.6.
The probability of Linda passing is 0.7.
Work out the probability of

(a) Steve not passing

(b) Linda not passing

(c) both Steve and Linda passing

(d) both Steve and Linda not passing

(e) Steve passing and Linda not passing

(f) Steve not passing and Linda passing.

> **Remember:**
> P(not A) = 1 − P(A)
> P(A and B) = P(A) × P(B)
> for independent events.

2 A bag contains 12 equal-sized coloured balls.
6 of the balls are red, 4 of the balls are blue and 2 of the balls are white.
A ball is selected at random, its colour is recorded and then the ball
is put back in the bag.
A second selection is then made at random. The colour of this ball is
also recorded.
Copy and complete the tree diagram.

> P(red) = $\dfrac{6}{12}$

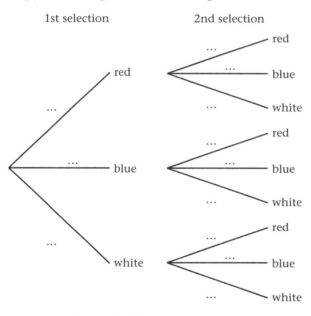

Work out the probability of the selected balls being

(a) both red

(b) both of the same colour

(c) one red and one blue

(d) of different colours •———

(e) at least one blue.

> P(both red) = $\dfrac{6}{12} \times \dfrac{6}{12}$

> You could select red and
> not red, or blue and not
> blue, or white and not
> white.

3 When Jaqui and Wendy play tennis the probability of Jaqui winning is $\frac{3}{4}$ and the probability of Wendy winning is $\frac{1}{4}$.

When Jaqui and Wendy play chess the probability of Jaqui winning is $\frac{1}{5}$ and, like tennis, one of the two girls **must** win – assume there is no chance of a game of chess being a draw.

Jaqui and Wendy play a game of tennis and a game of chess.

(a) Draw a probability tree diagram for this situation.

(b) Using your tree diagram, or otherwise, work out the probability of
  (i) Jaqui winning both games
  (ii) Wendy winning at least one of the games
  (iii) the girls winning one game each.

4 Nikki and Ramana both try to score a goal in netball.
The probability that Nikki will score a goal on her first try is 0.65.
The probability that Ramana will score a goal on her first try is 0.8.

(a) Work out the probability that Nikki and Ramana will both score a goal on their first tries.

(b) Work out the probability that neither Nikki nor Ramana will score a goal on their first tries. [E]

5 Just before a by-election 1500 people were asked which political party they intended to vote for.
The results of the survey were

Conservative	420
Labour	700
Lib. Dem.	300
Green Party	80

There were no other parties standing for election.
On the day of the election two people were stopped at random just before they cast their votes.
Work out, giving your reasons, the best estimate of the probability of

(a) these two people both voting Labour

(b) these two people both voting for the same party

(c) these two people voting for different parties.

6 Here are diagrams of two spinners.

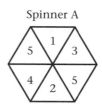 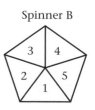

Spinner A   Spinner B

Naomi will spin each spinner once and record the numbers they land on.

(a) Work out the probability of these two numbers
   (i) both being 5
   (ii) both being the same
   (iii) having a sum of 4.

(b) Work out the probability of the difference between the numbers being
   (i) zero
   (ii) 1.

7 In darts the probability that Lucy will win is 0.7.
In pool the probability that Emma will win is 0.6.
The outcomes of the games are independent of each other.
In both games, if Lucy does not win then Emma wins.
The games cannot finish as a tie.

(a) Copy and complete the tree diagram.

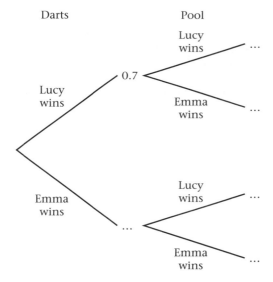

(b) Using the tree diagram, or otherwise, work out the probability that
   (i) Lucy will win both games
   (ii) Emma will win at least one game.

8 Sumreen and Asif are both about to take a music examination.
The probability that Sumreen will pass is 0.8.
The probability that Asif will pass is 0.75.
Work out the probability that

(a) they will both pass

(b) they will both fail

(c) at least one of them will pass.

9 Anna is due to take her driving test. The probability that she will pass at the first attempt is 0.6. If she fails at the first attempt, the probability of her passing at the second or any subsequent attempt is 0.8.

(a) Complete the probability tree diagram.

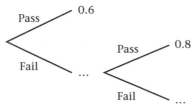

1st attempt   2nd attempt   3rd attempt

(b) Work out the probability of Anna passing the driving test
  (i) at the second attempt
  (ii) within, at most, two attempts
  (iii) within, at most, three attempts.

10 Jim is about to take a music test. The probability of him passing the test at the first attempt is p. If he fails, the probability of him passing on the second or any subsequent attempt is q.

Prove that the probability of Jim passing the test in no more than two attempts is

$$p + q - pq$$

25 Transformations of graphs

Exercise 25.1
Link 25A

1 $f(x) = 3x + 1$. Find
   (a) $f(2)$ (b) $f(7)$ (c) $f(0)$ (d) $f(-4)$ (e) $f(\frac{1}{4})$

2 $f(x) = 3x$. Find
   (a) $f(-x)$ (b) $f(3x)$ (c) $f(x + 1)$
   (d) $f(2x + 1)$ (e) $-f(x)$ (f) $-f(-x)$

3 $f(x) = x^3 - 1$. Find
   (a) $f(0)$ (b) $f(-1)$ (c) $f(3)$
   (d) $f(\frac{1}{2})$ (e) $f(2x)$ (f) $f(-x)$
   (g) $\frac{1}{2}f(x)$ (h) $f(1 - x)$

Exercise 25.2
Links 25B, C

Copy and complete.

1 The graph of $y = 2x + 3$ is the graph of $y = 2x$ translated ...

2 The graph of $y = 3x^2 - 2$ is the graph of $y = 3x^2$ translated ...

3 The graph of $y = \dfrac{1}{x} + 4$ is the graph of $y = \dfrac{1}{x}$ ...

4 The graph of $y = 3x + 4$ is the graph of ... translated 4 units
   vertically in the positive y-direction.

5 The graph of $y = x^2 + 5$ is the graph of $y = $ ... translated 3 units
   vertically in the positive y-direction.

6 The graph of $y = x^3 + 1$ is the graph of ...

> **Remember:** $y = f(x) + k$ moves the graph of $y = f(x)$ up by k units.

Exercise 25.3
Links 25D–F

1 Show how the following graphs are related to the graph of $y = x^2$.
   In each case give the coordinates of the vertex.
   (a) $y = x^2 + 2x + 1$ (b) $y = x^2 - 2x + 1$
   (c) $y = x^2 + 16x + 64$ (d) $y = x^2 + 5x + 6.25$

> Complete the square,
> e.g. $y = x^2 + 6x + 1$
> is $y = (x + 3)^2 - 8$

2 Copy and complete.

(a) $y = (x - 2)^2$ is the graph of $y = x^2$ translated ... units in the positive x-direction.

(b) $y = (x + 3)^3$ is the graph of $y = x^3$ translated ...

(c) $y = \dfrac{1}{(x + 2)}$ is the graph of $y = \dfrac{1}{x}$ ...

(d) $y = 3(x + 1)$ is the graph of $y = 3x$ ...

> **Remember:**
> $f(x - 2)$ is a translation of $\begin{pmatrix} 2 \\ 0 \end{pmatrix}$.

3 (a) $y = (x - 3)^3 + 1$ is the graph of $y = x^3$ translated 3 units ... and 1 unit ...

(b) $y = \dfrac{1}{(x - a)} + b$ is the graph of $y = \dfrac{1}{x}$ translated ...

(c) $y = (x + 2)^3 - c$ is the graph of $y = x^3$ translated ...

(d) $y = 2x + 5 = 2(x + 2) + 1$ is the graph of $y = 2x$ translated ...
$= 2(x + 1) + 3$ is the graph of $y = 2x$ translated ...
$= 2(x) + 5$ is the graph of $y = 2x$ translated ...

4 Sketch the graph of $y = x^2$.

(a) Use a tracing of this to sketch on the same axes the graph of $y = (x - 3)^2 + 1 = x^2 - 6x + 10$.

(b) Rearrange the equation $y = x^2 - 6x + 5$ as $y = (x - p)^2 + q$ in order to sketch the graph.

5 Relate the following graphs to $y = x^2$ using the method of question **4 (b)**.

In each case give the coordinates of the vertex.

(a) $y = x^2 + 6x + 7$

(b) $y = x^2 - 2x + 8$

(c) $y = x^2 + 5x + 0.25$

6 Sketch the curve $y = \dfrac{1}{x}$ for $x > 0$.

(a) Use a tracing of this to sketch on the same axes the graph of

$$y = \dfrac{1}{(x - 1)} + 2 = \dfrac{2x - 1}{x - 1}$$

(b) Sketch the graph of $y = \dfrac{1}{x + 3} - 1$.

Exercise 25.4

Links 25G–I

1 This is the graph of $y = f(x)$ where
$$f(x) = (x - 1)(x - 3)$$
$$= x^2 - 4x + 3$$

    (a) Find the equation of $f(-x)$.

    (b) Sketch the graph of $f(-x)$.

    (c) Sketch the graph of $-f(x)$ and write down the equation.

Mark the coordinates of all intercepts and vertices.

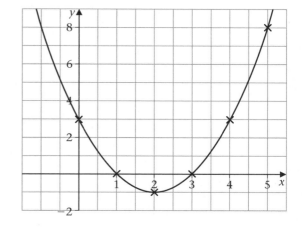

2 Make a flow chart which transforms $y = x^2$ into

$$y = -x^2 - 4x + 3$$

List the separate transformations and draw them on a graph.

> Work one step at a time.

> $-x^2 - 4x + 3$
> $= -(x + 2)^2 + 7$

3 Draw the flow chart which transforms $y = \dfrac{1}{x}$ into $y = \dfrac{-1}{x - 3} + 2$.

List the separate transformations in order.

4 Draw the flow chart which transforms $y = x^3$ into $y = x^3 - 3x^2 + 3x - 7$.
List the separate transformations in order.

5 List the transformations, in the correct order, which when applied to $y = x^2$ give the graphs of the following.

    (a) $y = x^2 + 6x - 2$ **(b)** $y = 4x - x^2$

    (c) $y = 2x^2 + 2x - \frac{1}{2}$ **(d)** $y = 4x^2 + 2$

    (e) $y = 1 - 9x^2$

Sketch each graph showing all intercepts and vertices.

6 List the transformations, in the correct order, which when applied to $y = \dfrac{1}{x}$ give the graphs of the following.

    (a) $y = \dfrac{1}{3x} - 1$ **(b)** $y = \dfrac{3}{x} - 1$

    (c) $y = 2 - \dfrac{4}{2 + x}$ **(d)** $\dfrac{-1}{2(x - 2)}$

Sketch each graph.

Exercise 25.5

Link 25J

1 For each of the graphs in the diagram, select which function represents it and say how it relates to the graph of the function sin x.

Remember:
In $y = A\sin(Bx + C) + D$
A is a vertical stretch
B changes the number of repeats
C moves left or right
D moves up or down

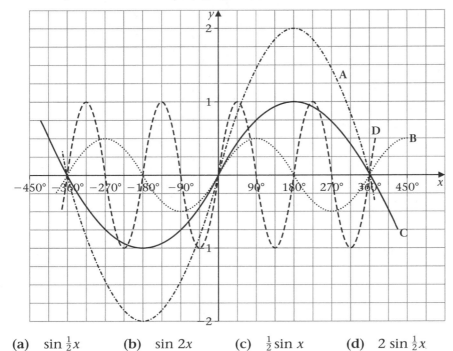

(a) $\sin\frac{1}{2}x$ **(b)** $\sin 2x$ **(c)** $\frac{1}{2}\sin x$ **(d)** $2\sin\frac{1}{2}x$

2 The diagram shows $f(x) = \tan x$.
Explain why $\tan x = \tan(x + 180°)$

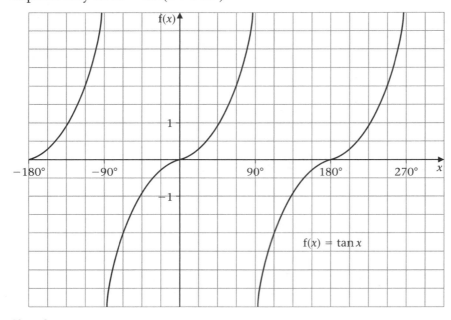

Sketch

(a) $\tan 2x$ **(b)** $\tan(x + 90°)$ **(c)** $2\tan\frac{1}{2}x$ **(d)** $\tan\frac{1}{2}(x + 90°)$

26 Circle theorems

Exercise 26.1 Link 26A

1 Calculate the marked angles in these diagrams. O is always the centre of the circle.

> **Remember:** The angle between a radius and a tangent is 90°.

(a)

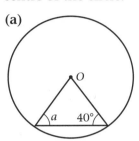

(b)

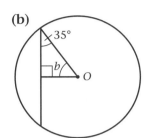

(c)

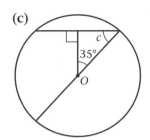

(d)

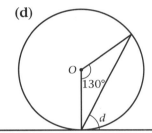

(e)

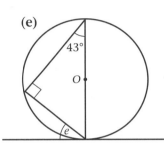

2

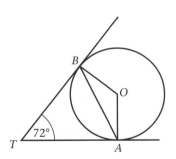

O is the centre of the circle. TA and TB are tangents to the circle.
Angle $ATB = 72°$.
Calculate the size of **(i)** $\angle BAT$
                      (ii) $\angle OBA$

> **Remember:** The lengths of two tangents from a point to a circle are equal.

3 Calculate the marked angles in these diagrams. O is always the centre of the circle. The lines drawn to the circles are tangents.

(a)

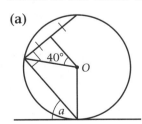

(b)

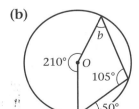

(c)

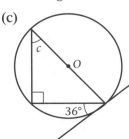

4 In the diagram, PQ is a chord of the circle centre O.
The radius of the circle is 17 cm.
M is the midpoint of PQ.
$PQ = 30$ cm.
Calculate the length of OM.

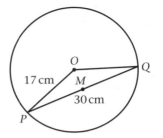

5 PQ and PR are tangents to the circle centre O.
The angle $QPR = 32°$.
Calculate the angle POR.

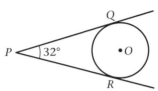

Exercise 26.2 Link 26B

1 AB is a diameter of the circle.
C is a point on the circumference of
the circle.
$AC = 5$ cm and $BC = 12$ cm.
Calculate

(a) the radius of the circle

(b) the area of the circle.

Remember: The angle
in a semicircle is a right
angle.

2 AC is a diameter of the circle centre O.
B is a point on the circumference of the circle.
P is another point on the circumference of the circle.
PO is parallel to AB.
The angle $ACB = x°$.
Find, with reasons and in terms of x, an expression for the angle
AOP.

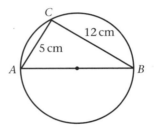

3 A, B, C and D are four points on the circumference of a circle.
$ABCD$ is a rectangle. The area of the circle, in terms of π, is
400π cm^2.
Given that the lengths of AD and DC are integers show that the
area of $ABCD$ is 768 cm^2.

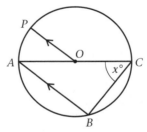

4 AB is a diameter of the circle.
The line STA is a tangent to the circle.
C is a point on the circumference of the
circle and is such that $TC = CA$.
The angle $CTS = x°$.
Find, in terms of x, an expression for the angle

(a) TCA (b) ABC

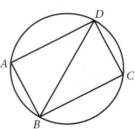

Hint: The triangle
is an enlargement
of a 3, 4, 5 triangle.

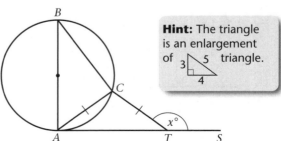

5 P, Q and R are three points on the circle centre O.
Angle $PQR = 54°$.
Find the angle

(a) POR (b) RPO

Give your reasons.

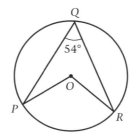

Remember: The angle subtended at the centre of a circle is twice the angle at the circumference.

6 P, Q, R and S are four points on the circle centre O.
The reflex angle $POR = 230°$.
Calculate the size of the angle

(a) PSR (b) PQR

Give your reasons.

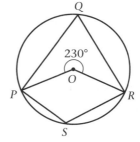

7 The four points A, B, C and D lie on the circle centre O.
Angle $BCD = 48°$.
$AB = AD$.
Calculate the size of the angle

(a) ADB (b) ODB

Give your reasons.

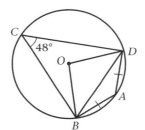

Remember: Opposite angles of a cyclic quadrilateral are supplementary.

8 P, Q, R and S are four points on the circumference of the circle centre O.
The line segment QOT is a diameter of the circle.
The angle $QRS = x°$.
$PS = PQ$.
Find, in terms of x, expressions for

(a) the angle QTS (b) the angle QOS

(c) the angle QPS (d) the angle SQP

(e) the angle TSO (f) the angle TOS.

In the case when the triangle TOS is **equilateral**

(g) find the value of x.

9 P, Q, R and S are four points on the circle centre O.
PR and QS intersect at the point M.
Angle $SPR = x°$ and angle $RMQ = y°$.
Find, in terms of x and/or y, expressions for the angles

(a) RQS (b) PSQ

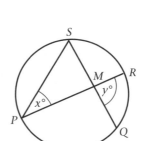

Remember: Angles in the same segment are equal.

10 P, Q, R and S are four points on a circle.
T is a point outside the circle with PST a straight line.
Angle PQR = 48°.
Calculate the size of the angle RST.

Give your reasons.

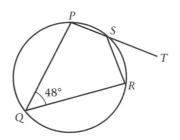

11 PQRS is a cyclic quadrilateral.
O is the centre of the circle.
The point T lies on the circle and QOT is a straight line.
The angle QTS = x°.
Write down, in terms of x, expressions for

(a) the angle QPS   (b) the angle QOS   (c) the angle QRS.

Give your reasons.

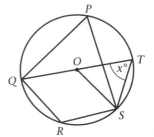

12 ABCD is a cyclic quadrilateral.
The line from D to the centre of the circle, O, is parallel to AB.
The angle DCB = 62°.
Calculate the angle

(a) DOB   (b) DAB   (c) OBA

Give your reasons.

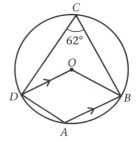

13 The diagram shows two circles which intersect at
the points Q and R.
PQRS and NMRQ are two cyclic quadrilaterals.
P, Q and N lie on a straight line.
Angle NRM = 25°, angle QNR = 55° and
angle QMN = 53°.
Calculate the size of the angle

(a) RNM   (b) MRQ   (c) PSR

Give your reasons.

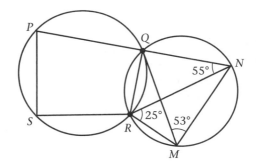

Exercise 26.3   **Link 26C**

1 AB and CD are two chords of a circle which
meet inside the circle at the point M.

(a) Prove that the triangles MAC and MDB
are similar.

(b) Show that MA × MB = MD × MC.

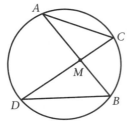

Remember: Angles in
the same segment are
equal.

2 *PQ* and *RS* are two chords of a circle which meet at a point *N* which lies outside the circle.

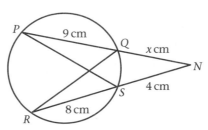

(a) Explain why the triangles *NPS* and *NRQ* are similar.

(b) Prove that
$NQ \times NP = NS \times NR$.

$NQ = x$ cm, $QP = 9$ cm, $NS = 4$ cm and $SR = 8$ cm.

(c) Show that $x^2 + 9x - 48 = 0$.

(d) Solve the equation to find the value of *x*.

3 The chord *AB* of the circle, when produced, meets the tangent at *C* at the point *T*.

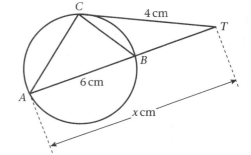

(a) Explain fully, why the triangles *TCA* and *TBC* are similar.

(b) Show that

(i) $\dfrac{TC}{TB} = \dfrac{TA}{TC}$ (ii) $TC^2 = TA \times TB$

$TA = x$ cm, $AB = 6$ cm and $TC = 4$ cm.

(c) Prove that $x^2 - 6x - 16 = 0$.

(d) Solve this equation to find the value of *x*.

4 *PA* and *PB* are tangents to a circle centre *O*, which meet at *P*. *PO* produced meets the circle at *C*.
Angle $APO = x°$.
Prove that
angle $ACP =$ angle $BCP = \frac{1}{2}(90 - x)°$.

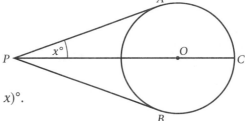

5 Two circles, centres *O* and *Q*, touch at *T*.
The common tangent at *T* meets the other common tangents *AB* and *CD* at *X* and *Y* respectively.

(a) Explain why *AOTX* is a cyclic quadrilateral.

(b) Explain why angle $BXT =$ angle AOT.

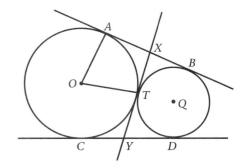

6 Two circles touch externally at a point M.
A line through M meets the circles at
points A and B.
Prove that the tangent at A and the
tangent at B **must** be parallel.

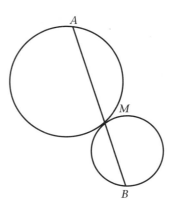

7 $ABCD$ is a cyclic quadrilateral. O is the centre of the circle.
$AB = BC$.

(a) Prove that angle $ADC = 2 \times$ angle ACB.

Angle $BAC = x°$.

(b) Write down an expression for the angle AOC.

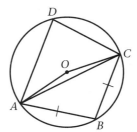

8 Chords AB and CD meet at the point P outside the circle.

(a) Write down, with reasons, a triangle similar to APD.

$PB = x$ cm, $BA = 4$ cm, $PD = 5$ cm and $DC = 7$ cm.

(b) Prove that $x^2 + 4x - 60 = 0$.

(c) Find the value of x.

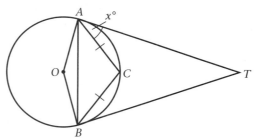

9 AT and BT are two tangents of
a circle centre O, which meet
at the point T.
$AC = BC$ and angle $TAC = x°$.

(a) Find, in terms of x,
expressions for
(i) angle ACB
(ii) angle BOA.

(b) Explain whether or not the
quadrilateral $AOBC$ can be cyclic.

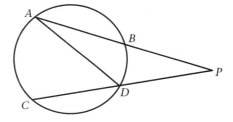

A quadrilateral can
only be cyclic if its
opposite angles are
supplementary.

10 A, B and C are points on a
circle, centre O.
TA is a tangent.
Angle $C\hat{A}T = x°$.
Without stating the alternate
segment theorem, prove that
angle $A\hat{B}C = x°$.

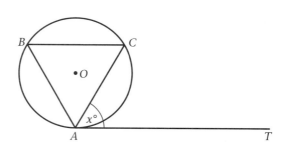

11 The diagram shows a circle centre *O*. *PQ* and *RQ* are
tangents to the circle at *P* and *R* respectively.
S is a point on the circle.
Angle *PSR* = 70°.
PS = *SR*.

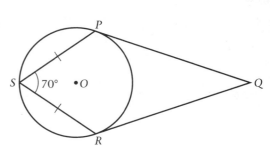

   (a) **(i)** Calculate the size of the angle *PQR*.
     (ii) State the reasons for your answer.

   (b) **(i)** Calculate the size of the angle *SPQ*.
     (ii) Explain why *PQRS* cannot be a cyclic quadrilateral.   [E]

12 *A*, *B* and *C* are three points on the circle.
Each of the angles in the triangle *ABC* is acute.
The tangents at *A* and *C* meet at *T*.
M is the point on *AT* such that angle *ACM* = angle *TCM*.
Angle *CMT* = 81°.
Giving all of your reasons, calculate the size of the angle *ABC*.

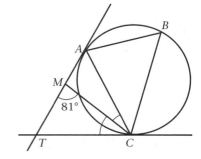

27 Vectors

Exercise 27.1

Links 27A–C

1 P is the point (2, 1), Q is the point (4, 5) and R is the point $(-2, 3)$. Write the column vectors

(a) $\overrightarrow{PQ}$ (b) $\overrightarrow{QR}$ (c) $\overrightarrow{RQ}$ (d) $\overrightarrow{PR}$

2 $\overrightarrow{CD}$ is the vector $\begin{pmatrix} 3 \\ -2 \end{pmatrix}$ and $\overrightarrow{DE}$ is the vector $\begin{pmatrix} 5 \\ -4 \end{pmatrix}$.

Find the vectors   (a) $\overrightarrow{CE}$ (b) $\overrightarrow{EC}$

3 $a = \begin{pmatrix} 2 \\ 3 \end{pmatrix}$, $b = \begin{pmatrix} 5 \\ -8 \end{pmatrix}$, $c = \begin{pmatrix} -5 \\ -3 \end{pmatrix}$.

(a) Write these as column vectors.
     (i) $2a$ (ii) $3a + 2c$ (iii) $5b - 3c$
     (iv) $4c + 3a$ (v) $4a - 2b$ (vi) $5b - 2a$
     (vii) $-3b$ (viii) $b - a$

(b) Find the vector d such that
     (i) $2a + d = c$ (ii) $d - b = 2c$

(c) Write down a vector that is parallel to $a + b + c$.

4 $p = \begin{pmatrix} 5 \\ 2 \end{pmatrix}$, $q = \begin{pmatrix} -2 \\ 4 \end{pmatrix}$, $r = \begin{pmatrix} -1 \\ -1 \end{pmatrix}$.

(a) Write these as column vectors.
     (i) $3p$ (ii) $2p + 3r$ (iii) $5q - 3r$
     (iv) $2r + 5p$ (v) $2p - 2q$ (vi) $7q - 2p$
     (vii) $-6q$ (viii) $q - p$

(b) Find the vector s such that
     (i) $3p + s = r$ (ii) $s - 3q = 5r$

(c) Write down a vector that is parallel to $p + q + r$ and twice the length of it.

Remember:
$a + b$

$a + b$ b   a

$a - b = a + (-b)$

a $a - b$ $-b$

Exercise 27.2

Links 27D–F

1 Find the magnitude of these vectors.

(a) $\begin{pmatrix} 4 \\ 3 \end{pmatrix}$ (b) $\begin{pmatrix} -6 \\ 8 \end{pmatrix}$ (c) $\begin{pmatrix} -1 \\ -1 \end{pmatrix}$

(d) $\begin{pmatrix} 5 \\ 12 \end{pmatrix}$ (e) $\begin{pmatrix} 2 \\ -2 \end{pmatrix}$

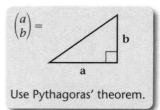

$\begin{pmatrix} a \\ b \end{pmatrix} =$

Use Pythagoras' theorem.

2 $\mathbf{p} = \begin{pmatrix} 3 \\ 4 \end{pmatrix}$, $\mathbf{q} = \begin{pmatrix} -5 \\ 12 \end{pmatrix}$, $\mathbf{r} = \begin{pmatrix} -1 \\ -1 \end{pmatrix}$.

Work out

(a) $|\mathbf{p}|$ (b) $|\mathbf{q}|$ (c) $|\mathbf{r}|$ (d) $|\mathbf{p} + \mathbf{q}|$

(e) $|\mathbf{p} - \mathbf{q}|$ (f) $2|\mathbf{r}|$ (g) $|3\mathbf{r}|$ (h) $|\mathbf{p} + \mathbf{r}|$

(i) $|\mathbf{q} - \mathbf{r}|$ (j) $|\mathbf{p} + \mathbf{q} - \mathbf{r}|$

> $|\mathbf{p}|$ means the magnitude of vector **p**.

3 $\mathbf{a} = \begin{pmatrix} 2 \\ 3 \end{pmatrix}$, $\mathbf{b} = \begin{pmatrix} -2 \\ 1 \end{pmatrix}$, $\mathbf{c} = \begin{pmatrix} -1 \\ -1 \end{pmatrix}$.

Calculate **x** when

(a) $\mathbf{a} + \mathbf{x} = \mathbf{b}$ (b) $2\mathbf{b} + \mathbf{x} = \mathbf{c}$ (c) $3\mathbf{c} + \mathbf{x} = 2\mathbf{a}$

(d) $\mathbf{x} - 4\mathbf{c} = \mathbf{a}$ (e) $3\mathbf{x} - \mathbf{a} = \mathbf{b}$

4 $\mathbf{p} = \begin{pmatrix} 2 \\ -1 \end{pmatrix}$, $\mathbf{q} = \begin{pmatrix} -2 \\ -3 \end{pmatrix}$, $\mathbf{r} = \begin{pmatrix} -1 \\ 0 \end{pmatrix}$.

(a) $2\mathbf{p} + a\mathbf{q}$ is parallel to the x-axis. Find the value of a.

(b) $3\mathbf{p} - b\mathbf{r}$ is parallel to the y-axis. Find the value of b.

> **Remember:** Vectors parallel to the x-axis are of the form $\begin{pmatrix} x \\ 0 \end{pmatrix}$.
>
> Vectors parallel to the y-axis are of the form $\begin{pmatrix} 0 \\ y \end{pmatrix}$.

5 $\mathbf{a} = \begin{pmatrix} 2 \\ 1 \end{pmatrix}$, $\mathbf{b} = \begin{pmatrix} -1 \\ 3 \end{pmatrix}$.

Find the values of p and q such that $p\mathbf{a} + q\mathbf{b} = \begin{pmatrix} 5 \\ -8 \end{pmatrix}$.

6 A is the point $(3, 2)$, O is the point $(0, 0)$ and the position of a variable point P is given by

$$\overrightarrow{OP} = \overrightarrow{OA} + t\begin{pmatrix} 1 \\ 1 \end{pmatrix}$$

Calculate the coordinates of P for integer values of t between -2 and $+4$. Write down the equation of the path of P, as t varies, in the form $y = mx + c$.

7 P is the point $(2, 1)$, Q is the point $(1, 3)$ and R is the point $(-2, -2)$.

(a) Write down the coordinates of the midpoint of

    (i) PQ (ii) QR (iii) PR

(b) S lies on PQ extended so that $PS = 3 \times PQ$. Work out the coordinates of S.

(c) T lies on PR so that $PT = \frac{1}{3}PR$. Work out the coordinates of T.

> **Remember:** For two vectors **a** and **b**, the position of the vector of the midpoint between them is $\frac{1}{2}(\mathbf{a} + \mathbf{b})$.

8 *ABCD* is a rhombus. *A* has coordinates (1, 2), *B* has coordinates (3, 5) and the coordinates of the point of intersection of the diagonals is (3, 2).

    (a) Find the position vectors of the points *C* and *D*.

    (b) Work out the vector of the line joining the midpoints of *AB* and *CD*.

Exercise 27.3   Link 27G

1 $\overrightarrow{AB} = \mathbf{a}$ and $\overrightarrow{AC} = \mathbf{c}$. *D* and *E* are the midpoints of *AB* and *AC* respectively.

    (a) Write down in terms of **a** and **c** the vectors

        (i) $\overrightarrow{AD}$ **(ii)** $\overrightarrow{AE}$ **(iii)** $\overrightarrow{DE}$ **(iv)** $\overrightarrow{BC}$

    (b) Write down the geometrical relationship between the lines *DE* and *BC*.

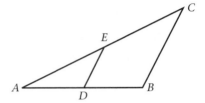

2 $\overrightarrow{AB} = \mathbf{b}$ and $\overrightarrow{AC} = \mathbf{c}$. *D* and *E* are $\frac{3}{4}$ of the way along *AB* and *AC* respectively. Find $\overrightarrow{BC}$ and $\overrightarrow{DE}$ in terms of **b** and **c** and write down the geometrical relationship between the lines *DE* and *BC*.

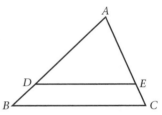

3 $\overrightarrow{AE} = \mathbf{e}$ and $\overrightarrow{AD} = \mathbf{d}$. $CE : AE = 2 : 1$ and $BD : AD = 2 : 1$.

Write the vectors $\overrightarrow{DE}$ and $\overrightarrow{BC}$ in terms of **e** and **d** and then write down two facts about the lines *DE* and *BC*.

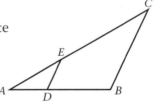

4 *ABCD* is a parallelogram, *P*, *Q*, *R* and *S* are the midpoints of *AB*, *BC*, *CD* and *DA* respectively.
$\overrightarrow{AB} = \mathbf{a}$ and $\overrightarrow{AD} = \mathbf{d}$.
Write $\overrightarrow{PQ}$, $\overrightarrow{QR}$, $\overrightarrow{SR}$ and $\overrightarrow{PS}$ in terms of **a** and **d**.
Write down the name of the quadrilateral *PQRS*.

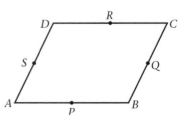

5 $\overrightarrow{AD} = \mathbf{x}$ and $\overrightarrow{AE} = \mathbf{y}$. $AC = 3AE$ and $AB = 3AD$.

    (a) **(i)** Find $\overrightarrow{DE}$ and $\overrightarrow{BC}$ in terms of **x** and **y**.
         (ii) Hence explain why triangles *PDE* and *PBC* are similar.

    (b) Express $\overrightarrow{DP}$ and $\overrightarrow{AP}$ in terms of **x** and **y**.

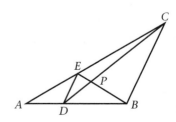

6 *ABCDEF* is a regular hexagon with its centre at point *O*.
$\overrightarrow{OB} = \mathbf{b}$ and $\overrightarrow{OC} = \mathbf{c}$.

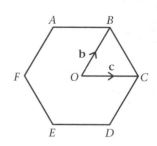

   (a) Write in terms of **b** and **c** the vectors

      (i) $\overrightarrow{BC}$ **(ii)** $\overrightarrow{CD}$ **(iii)** $\overrightarrow{DE}$ **(iv)** $\overrightarrow{EF}$

      (v) $\overrightarrow{FA}$ **(vi)** $\overrightarrow{AB}$ **(vii)** $\overrightarrow{AC}$ **(viii)** $\overrightarrow{FD}$

   (b) Prove that *ACDF* is a rectangle.

7 In triangle *ABC*, *M* is the midpoint of *BC* and *N* is the midpoint of *AC*. $\overrightarrow{AB} = \mathbf{b}$ and $\overrightarrow{BC} = \mathbf{c}$.
BN is produced to a point *D* so that $BD : BN = 4 : 3$.

   (a) Write $\overrightarrow{AM}$ and $\overrightarrow{DC}$ in terms of **b** and **c**.

   (b) Prove that *AM* is parallel to *DC* and that $AM : DC = 3 : 2$.

8 In triangle *ABC*, *P* is the point on *AB* so that $AP = \frac{2}{3}AB$ and *Q* is the midpoint of *BC*. *AC* is produced to a point *R* so that
$AC = CR$. $\overrightarrow{AB} = \mathbf{b}$ and $\overrightarrow{AC} = \mathbf{c}$.

   (a) State in terms of **b** and **c** the vectors

      (i) $\overrightarrow{AP}$ **(ii)** $\overrightarrow{AQ}$ **(iii)** $\overrightarrow{AR}$ **(iv)** $\overrightarrow{PQ}$ **(vi)** $\overrightarrow{QR}$

   (b) Show that *PQR* is a straight line and state the ratio of the lengths $PQ : QR$.

28 Introducing modelling

Exercise 28.1

Links 28A, B

1 George bought a new car on 1 February 2006 for £12 000.
The value of the car **depreciates** by 15% each year.

(a) Copy and complete this table for the value of the car, giving the values correct to the nearest £10.

Year	2006	2007	2008	2009	2010	2011
Value on 1 February (£)	12 000					

This is modelling using exponential functions of the form $y = a^x$, where a is a constant and x the variable.

(b) Draw a graph of the value of the car against the year from 2006 to 2011.

(c) Find a formula which models the value of the car n years after it was bought new.

(d) Use your formula to work out
  (i) the value of the car on 1 August 2016
  (ii) the year in which the value of the car first falls below £500.

2 Julie started to work for a jeweller on 1 January 2006.
Her wage was fixed at £250 per week.
The jeweller promised Julie a pay rise of 5% per year.
If Julie stays with the jeweller, what could be her expected weekly pay after she has been working there for

(a) 2 years   (b) 8 years   (c) n years?

3 t hours after midnight one day, the depth of water, d metres, at the entrance to a small fishing harbour is modelled by the formula

$$d = 8 + 5 \sin(20t)°$$

(a) Calculate the depth of water at
  (i) 2 am   (ii) 6 am   (iii) 10 am

(b) Work out the lowest value for the depth of the water at the entrance to the harbour.

(c) Work out the largest value of the depth of the water at the entrance to the harbour.

(d) Find the times of low tide and high tide during the day.

(e) Sketch a graph of the depth of the water at the entrance to the harbour as the time varies from 0 to 24 hours.

4 The diagram represents a moving particle P and a fixed point O.

O   P

The particle moves in a straight line such that its distance from O, y metres, at any time t seconds is given by the formula

$$y = a + b\cos(30t)°$$

When $t = 0$, $y = 3$.
When $t = 2$, $y = 4$.

(a) Show that

$$a + b = 3$$
$$2a + b = 8$$

(b) Solve these equations to find the values of a and b.

(c) Sketch the graph of y against t for values of t from 0 to 10.

(d) Find the value of y when
(i) $t = 2$ (ii) $t = 5$

(e) Find the maximum distance between O and P and the values of t when this occurs.

Exercise 28.2   Links 28C, D

1 The table below contains information about the diameters, in centimetres, and heights, in metres, of ten horse chestnut trees.

Diameter (cm)	120	140	141	132	134	130	112	150	147	122
Height (m)	27	33	31	30	31	32	28	33	31	30

(a) Plot these points on a scatter diagram.

(b) Draw a line of best fit on your scatter diagram.

(c) Find the equation of your line of best fit.

(d) Use the equation of the line to work out estimates of
(i)   the height of a horse chestnut tree of diameter 135 cm
(ii)   the diameter of a horse chestnut tree of height 29 m.

> You could use any two points on your line of best fit to help find the equation of your line of best fit.

2 Repeat parts (a), (b) and (c) of question 1 for the heights and diameters of these beech trees.

Diameter (cm)	170	160	183	162	164	180	170	172	194	203	210	201
Height (m)	36	32	39	32	33	37	32	34	37	39	41	40

3 In an experiment, two variables x and y are thought to be connected by a relationship of the type

$$y = ax^2 + b$$

where a and b are constants.
Corresponding values of x and y are given in the table below.

x	1	1.5	2	2.3	3.5	4
y	-1	1.5	5	7.58	21.5	29

(a) By plotting a suitable straight line graph, show that x and y are related by the given type of relationship.

(b) Use your graph to work out the values of a and b.

Compare $y = ax^2 + b$ with $Y = mX + c$.

(c) Work out the values of
(i) x when $y = 3$ (ii) y when $x = 2.4$ (iii) x when $y = 19.5$

4 The table shows the stopping distance, d feet, for a car travelling at a speed of s mph.

s	20	30	40	50	60	70
d	40	75	120	175	240	315

(a) Copy and complete this table of values for $\dfrac{d}{s}$ against s.

s	20	30	40	50	60	70
$\dfrac{d}{s}$	2	2.5				

(b) Plot the graph of $\dfrac{d}{s}$ against s for values of s from 20 to 70.

(c) Explain fully why this graph confirms that d and s are connected by a relationship of the type

$$d = as^2 + bs$$

in which a and b are constants.

(d) Use your graph to work out the values of a and b.

Compare $d = as^2 + bs$ with $Y = mX + c$.

(e) Without using your graph, work out the value of
(i) the stopping distance for a car travelling at a speed of 35 mph
(ii) the speed a car was travelling at when it is known that its stopping distance was 300 feet
(iii) the stopping distance for a car travelling at 100 mph.

At the scene of an accident, skid marks indicate that a particular car stopped in a distance of 145 feet.
The maximum speed limit on the road is 45 mph.

(f) Explain whether or not the skid marks provide evidence that the car was breaking the speed limit just prior to the accident.

5 Shivana has been conducting an experiment in science.
She believes that two variables y and t are connected by a formula
of the type

$$y = at^2 + bt$$

She has five sets of results from her experiment.
These results appear in the table below.

t	1	2	3	4	5
y	5.1	8.3	9.3	8.1	5.1

(a) By drawing a suitable straight line graph, show that, to
within experimental error, the results confirm Shivana's
believed formula.

(b) Use your graph to find the values of a and b.

Shivana will test her formula by repeating the experiment for a
sixth time using $t = 4.5$.

(c) Use the formula to work out an approximate predicted value
for y in this case.

Exercise 28.3   Link 28E

1 This sketch shows part of the graph of

$$y = pq^x$$

It is known that the points $(0, 3)$, $(2, k)$ and $(4, 1875)$ lie on the
curve.
Use the sketch and the given information to find the values of
p, q and k.

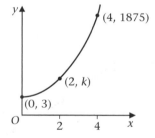

2 The point $(1, 16)$ lies on the curve $y = a^{2x}$.
Calculate the values of a.

3 The diagram represents a sketch of part of the graph of

$$y = p + q^x$$

Use the sketch to find the values of p, q and k.

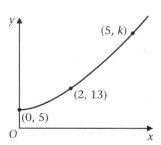

4 The time, t seconds, for one complete oscillation of a pendulum
of length l metres is given by the formula

$$t = kl^n$$

When $l = 1$, $t = 2$ and when $l = 9$, $t = 6$.
Work out the values of n and k.

29 Conditional probability

1   A bag contains 15 chocolates.
    8 of the chocolates are plain, 4 are milk and 3 are white.
    Joan selects a chocolate at random and eats it.
    Paul then selects a chocolate at random and eats it.

   (a)   Copy and complete the probability tree diagram.

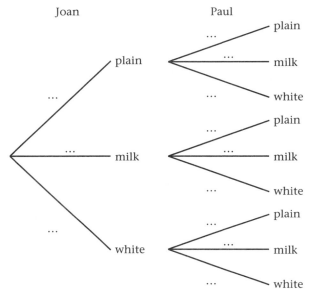

Joan   Paul

   (b)   Using the tree diagram, or otherwise, work out the
           probability of the two chocolates that are eaten being
           (i)   both plain
           (ii) of the same type
           (iii) of different types.

   (c)   Work out the probability of at least one of the chocolates
           that are eaten being white.

2   The 26 letters of the alphabet are placed in a bag.
    Two letters are selected at random from the bag.
    Find the probability of both of these selected letters being vowels.

> Select the letters one at a time.

3   A box contains 36 bags of crisps.
    15 of the bags contain plain crisps, 12 of the bags contain cheese
    and onion crisps and the remainder of the bags contain smoky
    bacon crisps.
    On a dark night, Jon selects 3 bags of crisps from the box.
    By using a tree diagram, or otherwise, work out the probability of
    the three bags of crisps being
   (a)   all of the same type   **(b)**   all of different types.

4   A bag contains 20 equal-sized counters of different colours.
Twelve are blue, five are red and three are white.
Fatima selects one of the counters at random and does not put it
back in the bag.
Collette then selects a second counter at random and she does not
put this counter back in the bag.

   (a)   Copy and complete the probability tree diagram for these
   selections.

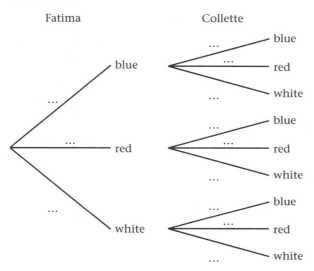

   (b)   Using the tree diagram, or otherwise, work out the
   probability of

   (i)   both selected counters being blue

   (ii)   both selected counters being the same colour

   (iii) at least one of the counters being white

   (iv)   neither of the selected counters being white.

Exercise 29.2   **Link 29C**

1   Asif travels to work by train.
The probability of the train being on
time is 0.8.
The train is never early.
If the train is on time the probability
of Asif getting on the train is 0.85.
If the train is late the probability of
Asif getting on it is 0.9.

   (a)   Draw a probability tree diagram for this situation.

   (b)   Use the diagram to work out the probability of Asif getting
   on the train on any particular day.

2 Mr Jones calculates that if a student regularly completes their homework the probability that they will pass the examination is 0.9 and that if the student does not do the homework the probability of their passing is only 0.45. Given that only 80% of his students do their homework regularly, calculating the probability that a student selected at random

(a) does not do the homework regularly and passes the examination

(b) passes the examination.

3 Carl is throwing darts at a dartboard.
He is trying to get all three darts in the bull's-eye.
The probability of his first dart landing in the bull's-eye is 20%.
If the first dart lands in the bull's-eye then the probability of his second dart landing in the bull's-eye is only 10%.
If the first dart does not land in the bull's-eye then the probability of the second dart landing in the bull's-eye is again 20%.
The probability of his third dart landing in the bull's-eye is affected by the number of darts already in the bull's-eye.
If there are no darts in the bull's-eye then the probability of the third dart landing in the bull's-eye is 20%.

Remember: A probability can be written as a fraction, percentage or decimal.

If there is one dart in the bull's-eye then the probability of the third dart landing in the bull's-eye is 10%.
If there are two darts in the bull's-eye then the probability of the third dart landing in the bull's-eye is only 5%.

(a) Copy and complete the probability tree diagram.

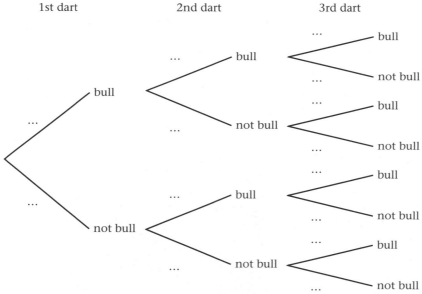

Carl throws three darts in succession at the bull's-eye.

(b) Work out the probability of
   (i) all three darts missing the bull's-eye
   (ii) all three darts landing in the bull's-eye
   (iii) at least one of the darts landing in the bull's-eye
   (iv) exactly one dart landing in the bull's-eye.

4 There are 4 red balls, 5 blue balls and 3 green balls in a bag.
A ball is to be taken at random and not replaced.
A second ball is then to be taken at random.

(a) Copy and complete the tree diagram.

(b) Use the tree diagram to calculate the probability that both balls taken will be
(i) red (ii) the same colour.

(c) Calculate the probability that exactly one of the balls taken will be red.

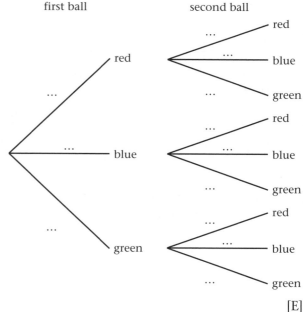

[E]

5 The probability of Steve passing his driving test at the first attempt is 0.7. If he fails at the first attempt then the probability of him passing at any subsequent attempt is 0.9.
By using a tree diagram, or otherwise, work out the probability of Steve passing his driving test

(a) at the second attempt

(b) in no more than two attempts.

> **Remember:**
> $P(\text{not } A) = 1 - P(A)$
> $P(A \text{ or } B) = P(A) + P(B)$
> $P(A \text{ and } B) = P(A) \times P(B)$

6 The probability of a family having a TV is 0.95.
If they have a TV the probability of the family having a freezer is 0.8.
If they do not have a TV the probability of the family having a freezer is 0.15.

(a) Copy and complete the tree diagram.

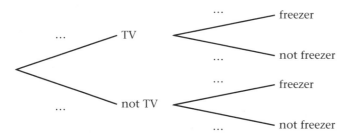

A family is selected at random.

(b) What is the probability of this family having
(i) a TV and a freezer
(ii) only a TV
(iii) only a freezer
(iv) neither a TV nor a freezer?

7 In her garden Mrs Mohammed has twelve fruit trees.
Six of the trees are apple, four are pear and the other two are plum.
During a storm, lightning strikes her trees twice.
Given that lightning **never** strikes the same thing twice, work out the probability that the two trees struck by lightning will

Read through the whole question before you start to draw the tree diagram because the choices to be made at each stage do not always become clear until the end.

(a) both be apple

(b) both be of the same type

(c) be of different types.